Mediators, Contract Men, and Colonial Capital

Rochester Studies in African History and the Diaspora

Toyin Falola, Series Editor
The Jacob and Frances Sanger Mossiker Chair in the
Humanities and University Distinguished Teaching Professor
University of Texas at Austin

Recent Titles

Ira Aldridge: The Last Years, 1855–1867
Bernth Lindfors

Population, Tradition, and Environmental Control in Colonial Kenya
Martin S. Shanguhyia

Humor, Silence, and Civil Society in Nigeria
Ebenezer Obadare

Nation as Grand Narrative: The Nigerian Press and the Politics of Meaning
Wale Adebanwi

The Rise and Demise of Slavery and the Slave Trade in the Atlantic World
Edited by Philip Misevich and Kristin Mann

The Power of the Oath: Mau Mau Nationalism in Kenya, 1952–1960
Mickie Mwanzia Koster

Cotton and Race across the Atlantic: Britain, Africa, and America, 1900–1920
Jonathan E. Robins

*Islam, Power, and Dependency in the Gambia River Basin:
The Politics of Land Control, 1790–1940*
Assan Sarr

Living Salvation in the East African Revival in Uganda
Jason Bruner

On Durban's Docks: Zulu Workers, Rural Households, Global Labor
Ralph Callebert

A complete list of titles in the Rochester Studies in African History and the Diaspora series may be found on our website, www.urpress.com.

Mediators, Contract Men, and Colonial Capital

Mechanized Gold Mining in the Gold Coast Colony, 1879–1909

Cassandra Mark-Thiesen

UNIVERSITY OF ROCHESTER PRESS

Copyright © 2018 by Cassandra Mark-Thiesen

All rights reserved. Except as permitted under current legislation, no part of this work may be photocopied, stored in a retrieval system, published, performed in public, adapted, broadcast, transmitted, recorded, or reproduced in any form or by any means, without the prior permission of the copyright owner.

First published 2018

University of Rochester Press
668 Mt. Hope Avenue, Rochester, NY 14620, USA
www.urpress.com
and Boydell & Brewer Limited
PO Box 9, Woodbridge, Suffolk IP12 3DF, UK
www.boydellandbrewer.com

ISBN-13: 978-1-58046-918-0
ISSN: 1092-5228

Library of Congress Cataloging-in-Publication Data

Names: Mark-Thiesen, Cassandra, author.
Title: Mediators, contract men, and colonial capital : mechanized gold mining in the Gold Coast Colony, 1879–1909 / Cassandra Mark-Thiesen.
Other titles: Rochester studies in African history and the diaspora ; v. 77.
Description: Rochester : University of Rochester Press, 2018. | Series: Rochester studies in African history and the diaspora ; v. 77 | Includes bibliographical references and index.
Identifiers: LCCN 2017050185 | ISBN 9781580469180 (hardcover : alk. paper)
Subjects: LCSH: Gold miners—Recruiting—Ghana—History. | Gold miners—Effect of technological innovations on—Ghana—History. | Labor supply—Ghana—History. | Industrial relations—Ghana—History. | Great Britain—Colonies—Africa—Administration—History. | Ghana—History—To 1957.
Classification: LCC HD8832 .M37 2018 | DDC 331.0422342209667090034—dc23 LC record available at https://lccn.loc.gov/2017050185

To my grandparents
and other teachers, farmers, housekeepers, and labor organizers

Contents

List of Illustrations	ix
Acknowledgments	xi
Introduction	1
1 Prospectors, Politicians, and the Question of "Progress": The First and Second Gold Booms in Wassa	16
2 Labor Recruitment in the Nineteenth Century: The Place of Practicality	52
3 Disrupted Recruitment at the Turn of the Twentieth Century: Women, Whites, and Other Labor Agents	101
4 Government Strategies for Assisting the Mines	119
5 Labor Agents, Chiefs and Officials, 1905–1909: The Incorporation of the Northern Territories' Labor Reserve	145
Conclusion	158
Notes	165
Bibliography	201
Index	209

Illustrations

Figures

1.1	The government railway to Kumase and Tarkwa, ca. 1900–1904	35
1.2	Managers and mineworkers at the Apinto mines in Wassa, ca. 1895–97	39
1.3	African miner standing in front of a mine shaft in the Gold Coast Colony, ca. 1900–1904	49
2.1	African surface miners in the Gold Coast Colony, ca. 1900–1904	55
2.2	Kru labor agent with other workers in the southern Gold Coast Colony, ca. 1876–77	78
2.3	A gang of Kru workers in the 1890s	82

Maps

1.1	The Protectorate of the Gold Coast, 1874–1919	17
1.2	The Tarkwa district showing concessions in 1906	50
3.1	French map of Liberia from 1835 with a large marker of the Kru Settlement on the southern Coast	104
3.2	West Africa, 1902	105

Tables

1.1	Yearly output of gold from 1886 to 1905 in the Southwest Gold Coast and Asante Region	36
1.2	Percy Tarbutt's business activities in Wassa	43
1.3	Active mining companies in the Gold Coast colony in 1904	51
2.1	Persons employed at Gold Coast gold mines, 1898	61
2.2	Labor in the mining industry of the Gold Coast and Asante, 1905	61

2.3	Labor employed in the mines, 1900–1918	74
2.4	Average daily wages paid at Abbontiakoon mines, ca. 1903	84
2.5	Daily wages paid to laborers in the mines in the Gold Coast and Asante, 1904	84
3.1	Labor in the mines in the Gold Coast and Asante, 1904 (by social origin)	107
4.1	Pay of gang leaders hired by the Government Transport Office (varying by experience), 1903	129

Acknowledgments

In the course of carrying out the underlying research for this book, I was able visit various parts of Ghana and the United Kingdom. I am thankful to the Janggen-Pöhn Foundation in St. Gallen (Switzerland), the Beit Fund of Oxford University, and St. Cross College for enabling this research. I also changed institutions and am grateful for the support of colleagues at both Oxford University and the University of Basel, especially the late Jan-Georg Deutsch and the late Patrick Harries.

Parts of this book have been presented in a number of forums: Gold Rush Imperialism Conference, Rothermere American Institute and Oxford Centre for Global History; Workshop on Wage Labour, Capital and Precarity in Global History and Africa, International Institute of Social History and African Studies Centre, Leiden; Roundtable on Ethnicity as a Political Resource: Perspectives from Africa, Latin America, Asia and Europe, University of Cologne; re: work the Summer Academy of re: work (or the IGK Work and Human Lifecycle in Global History at Humboldt University Berlin) at State University of Campinas; Tenth International Conference on Labour History, V. V. Giri National Labour Institute, Noida, India; International Conference on Forced Labour, Ho, Ghana; Research Seminar in African History, University of Basel; Masterclass of the International Institute of Social History, Amsterdam; Fourth European Conference on African Studies, Nordic Africa Institute, Uppsala; African Economic History Workshop, Geneva; Graduate Workshop in Economic and Social History, University of Oxford; and St. Cross Colloquia, University of Oxford. Many thanks to the organizers of these events and to members of the audience for their indispensable feedback.

I discovered source materials used in this book with the help of archivists and members of staff at a number of archives: the Ghana National Archives in Accra and Regional Archives Sekondi; archives of the University of Ghana, Legon; the Institute of African Studies at the University of Ghana; University of Mines and Technology, Tarkwa; Rhodes House Library; Cambridge University Archives; Public Record Office, London; London Metropolitan Archives; archives of the School of Oriental and African Studies; British Library Newspapers archive, Colindale; and the Basel Mission Archives.

Thanks also go out to those who have accompanied and encouraged me along the way, especially Peter Alexander, Gareth Austin, Kofi Baku, Rana Behal, Benjamin Brühwiler, Sarah Kunkel, Rouven Kunstmann, Enrique Martino, Deborah Mason, Heike Schmidt, and Julia Tischler. At the University of Rochester Press, I wish to thank the peer reviewers of my manuscript, as well as Toyin Falola and Sonia Kane and her colleagues, who were great to work with.

Special thanks go out to the members of my family, who make it worthwhile, and without whom this book would never have materialized: including my Edith, Yves, and Alex; Althea, Emmanuel, and Michael in Basel; Malaika, Carl, Zoë, and Miles in Tallahassee; Peter and Michaela in Hamburg; Tamara and Jürgen in Bad Oldesloe; Arlene W. in Sacramento; and Shirley M. in London—the list goes on.

Introduction

By 1910 the port of Sekondi in the western protectorate of the Gold Coast had already been serving as an entrepôt between Wassa and the expansive Atlantic world for several centuries. Dutch officials initially took an interest in the location because it possessed a good landing beach that was fairly well sheltered from strong winds and because of its close proximity to Wassa, which was known to be a rich source of gold. Although initially only land and water trade routes led to the gold-producing interior region, under the British colonial administration a government railway was constructed, reaching Tarkwa[1] in 1901, Obuase in 1902, Kumase in 1903, and Prestea in 1911. Unlike the cocoa and palm oil trade that determined the success of colonial ports in the eastern protectorate, to the west, the gold trade stood above all else, putting Sekondi in a dominant position as neighboring ports fell into a state of decay. It was a place of comings and goings, not just in trade (other primary exports included timber and rubber) but also in people. When the mechanized gold mines were established in the 1870s, miners from Britain, Australia, and the United States landed there in large numbers by way of a surfboat on their way to Wassa, as did most Liberian contract workers, for whom it was safer to travel along the coast by sea. By the turn of the century, Sekondi began to attract a growing number of migrant workers traveling on foot as well. Although some undertook the risk of a solo voyage, most of them arrived in Sekondi as part of a small gang led by an African labor agent. Sekondi was where agents serving the gold mines usually came to engage new recruits for contract work on the concessions or other types of wage work.

But if a port is the greatest signifier of a region's integration into the world economy, then the story that Sekondi told was one of haphazard incorporation. Although British officials had clearly invested time and money in the port city, its potential had simultaneously been tempered by mediocre expectations and missed opportunities. After a brief stay here during her travels through West Africa, the novelist Mary Gaunt described "a pretty place" straggling "up and down many hills."[2] Yet she could not help but notice how Sekondi's landscape was marked by a lack of planning and follow-through on the part of the British colonial administration, the port itself having only an "open roadstead" with "no harbor."[3]

The colonial state maintained a very lukewarm interest in the economic potential of the western protectorate, making it that much more puzzling to some observers when the railway was built from this southern port to stations in the mining districts to the north at the turn of the twentieth century. As it happened Britain's decision to build a railway from the port of Sekondi northward constituted a major turning point in the creation of a modern mining sector.[4] And yet, political friction between mining entrepreneurs and colonial officials continued to resurface. Sekondi's undone landscape aptly reflected their relationship during the early decades of mechanized gold mining. *Mediators, Contract Men, and Colonial Capital* is a social and economic history of West African miners in West Africa's first mechanized mining sector, which developed rapidly in the context of the colonial state's laissez-faire economic policy. The tenuous and shifting relationship between state officials and entrepreneurs in the gold mines in West Africa, which contrasted dramatically with what was occurring on white settler frontiers, is a key part of the investigation, especially in consideration of the opportunities it created for African entrepreneurs in and around this emergent wage-labor market. As Raymond Dumett put forward in 2012, "in West Africa the indigenous populations played a much larger role as active participants in the mining process than did their counterparts in the North American West, both as entrepreneurs and company promoters as well as wage laborers, in the supposedly European-led gold rushes, as traditional peasant miners and as members of the labor force."[5] In 1904, African employees in the Gold Coast's gold mining sector outnumbered European mining men by 17,044 to 611.

Mediators, Contract Men, and Colonial Capital takes on an explicitly labor-centered approach to this history by focusing on the evolution and persistence of indirect labor between 1879 and 1909. It reconstructs the emerging colonial economic context, illuminating how human geography and the preexisting social conditions of work in West Africa shaped capitalist transformation during this period. It carefully highlights the dynamics of the highly fluid West African labor market during this early stage of colonialism, while still taking into account how historical, social, and economic particularities may have shaped how different social groups responded to the opportunity of regular and extended wage work in the mines. The multitude of ways in which male, and some female, laborers from the Gold Coast and its protectorates as well as Liberia, Sierra Leone, and other parts of the region engaged with the new colonial economy with its groundbreaking technology, modern transportation services, easy access to credit, regular availability of jobs, and reinvented political institutions will be a key part of the discussion.

The Wassa gold mines were far from being the model of an imperial scheme. Therefore, the contradictions and inconsistencies of what some would describe as a new state of economic modernity are also highlighted. In particular, due to the pervasiveness of indirect recruitment, much of the politics of labor exploitation remained vested in African individuals. Labor agents were crucial mediators in West African mining labor markets. Yet in Wassa these actors were not necessarily representatives of individual mines or chiefdoms, either. Many exhibited inconsistent loyalties (at best). These were marginal character with a pure economic drive in their interactions with the mines. These recruiters are a neglected group in the history of gold mining in West Africa. A history of the processes of labor mobilization for the Wassa mines necessitates a story of the interplay of social, economic, and political transformations at the local, regional, colonial, and imperial level.

Wassa Gold Mining: A Long Story

Gold production and the gold trade from the region of Wassa flourished for centuries before the European presence. Although it is not impossible to distinguish when the mineral was first uncovered in the region, there is research to show that gold from the Upper Volta region was one of the three main sources for the trans-Saharan gold trade from at least 1400 on. Knowledge of this form of extra-subsistence activity eventually traveled south to the southern Akan region in medieval times, supposedly by means of the migration of Mandingo people. The Portuguese were the first Europeans to establish a direct trade in gold with peoples from this region in the 1490s, although the location of the mines and means of production were shrouded in secrecy. The first European sources listing the states of Wassa, Denkyera, Akyem, Asante, and Gyaman as the sources of Akan gold stemmed from the eighteenth century. The entire region, roughly 1.5 times the size of Switzerland, begins approximately 54 kilometers inland from the southwestern coast and extends to the northeast for about 220 kilometers in a rectangular shape, through the center of the forest zone to the Kwahu Escarpment. Previous research by Dumett has highlighted indigenous small-scale

4 *Introduction*

and seasonal mining production in these areas, based on family and some slave labor.[6] Dumett has also shown that this traditional form of mining continued to be a leading revenue generator for Africans even after the establishment of the first mechanized mines, expanding as it did due to new sources of transportation and trade.[7] Indeed, a major conclusion of his 1998 work, *El Dorado in West Africa*, was that small-scale mining outperformed the mechanized mines in terms of the amount of gold exported during the first gold rush of 1879–85.

Labor Transformations and Colonial Commerce in African History

This study builds on and expands prior historical research examining how colonial commerce promoted the transformation of labor regimes on the African continent. As it periodizes the expansion of the wage-labor market, it challenges some of the prevailing theoretical and conceptual models that scholars have applied to the history of colonial mining, often the foremost symbol of capitalism in Africa during this period. In expectation of an industrial revolution, historians flocked to the study of mining labor in Africa, which already was gaining popularity in the early 1950s. The mines comprised a familiar work environment in which historians anticipated witnessing Africans contribute to the rise of the so-called growth economy. By the end of the nineteenth century, mechanized mining in the western protectorate of the Gold Coast had managed to recruit thousands (in the double digits) of African workers. On the face of it, this was a notable achievement when we consider the legal end of slavery in the Gold Coast in 1874. Many scholars interpreted such growth in the wage-labor market as the first step in an inevitable process of expansion through which African workers could become proletarianized.[8] As a result, they focused on locating features of class consciousness. Thus, important information detailing the ways in which the work and life cycles of the African worker diverged from those of the Eurocentric model of the ideal worker was overlooked.

From the very start, however, certain scholars were not convinced by the thoughts surrounding class formation. Whereas some pointed to the lingering financial ties between the miners and their family members back home, others looked toward the social linkages taking shape in the mines themselves. Studies of mining in the Gold Coast have concentrated on the issue of ethnic organization in the mines, perceiving it as an extension of colonial policies. Although some social groups had no history of ethnicity prior to the colonial era,[9] and others certainly did show similar, albeit more fluid, forms of association, they all took on an ethnic identities in this workplace. Nevertheless, although scholars generally explained the social organization of African populations during the colonial era by way of ethnicity, they had to come to terms with how and why such allegiances were reinforced from below, by the miners themselves. This study further historicizes the transformation of ethnic identity during the colonial period by demonstrating the temporary nature of some of these unions in the late nineteenth- and early twentieth-century mining sector, though they were necessary for navigating the social landscape of the mines while the miners resided in Wassa during a roughly six-month period.

The Colonial State, African Authorities, and Labor Mobilization

This study has benefited from the increased caution taken by historians when making assumptions surrounding the relationships among the state, capital, and labor in the mines. Many historians of southern Africa have argued that the international finance networks of the mines were bolstered by their power in local government.[10] One important implication of such collusion was seen in the array of strategies the state used to mobilize labor for private firms, in addition to the men engaged to further its own expansionist ambitions.[11] As this and other studies show, the colonial government in the Gold Coast Colony also experimented with the recruitment of laborers for the gold mines.[12] At the same time, this study also illuminates the history of F. W. H. Migeod, the chief officer of the government transport office, and in particular disagreements between him and officials in Accra on the labor question, to forgo a monolithic view of the colonial state. On the whole, administrators were reluctant to identify with the Wassa mines in a formal and consistent manner.

Anne Phillips has shown that colonial economic priorities were more ambiguous in West Africa, due to the fundamental weaknesses of the colonial state.[13] There are, for example, few data backing the argument that colonial administrators intended to create an economic context favorable to the recruitment efforts of merchant capital on the West African coast when they made indigenous slavery illegal, as contended by Kenneth Swindell and Alieu Jeng.[14] Evasive behavior on the part of most colonial officials in the Gold Coast Colony in their dealings with the colonial mines in Wassa strengthens the argument that an ideological shift primarily informed the choice in favor of abolition. Even though enslaved Africans' access to work for missionaries, industries of scale, and the administration did add fire to the burning house of indigenous slavery, it was primarily the humanitarian ideals of missionaries, antislavery organizations and the public in the metropole that pushed the colonial state to pass the Emancipation Proclamation of 1874. The tenacity of enslaved men, women, and children forced colonial administrators to actually enforce the bill.

Nevertheless, most scholars have come to subscribe to the notion that the success of colonial commerce depended on one form or another of colonial intervention. Thus, the power of the colonial state cannot be ignored altogether, not even in non-settler colonies. Here, historians have given particular attention to the system of indirect rule in their discussion of labor recruitment for private firms[15] and government services[16] and the political constraints of African workers. Coercion has been primarily located in the combination of capitalist power and the persistence of social relations, which the state would not, and most likely could not, dismantle and managed to tap into through a variation of strategies. Such studies have shone a light on the colonial state's deep reliance on indigenous authorities, and especially the chiefs, who could mobilize their subjects in smaller or larger numbers.[17] However, in finding explanations for how African authorities were able to access the labor power of African workers during the formative years of colonial rule, most scholars have drawn on cultural history. Unfortunately, this highlighting of the exploitative power of the chiefs in the bulk of the literature leaves readers with an overwhelming sense that coercion by means of sociocultural power was the most pervasive tool of labor mobilization during the early colonial period. And these persistent references to social power alone run the danger of eliminating evidence of Africans as rational economic actors. Despite wide agreement that precolonial labor in West Africa was an economic as well as a social phenomenon, many labor historians of West Africa have only been writing sociocultural power into the history books.

Studies on the Wassa mines have tended to be more differentiated in their portrayal of indigenous authorities.[18] Scholars have, for instance, acknowledged the protective function of the chieftaincies where land rights were concerned. Others have contended that chiefs may not have had a great amount of say in responding to the labor demands of the state, nor did they draw great profits from such activities.[19] For the most part, this study breaks the focus on the chiefs by illuminating a moment in which an array of coexisting labor agents were attached to the mines. Moreover, it explores the economic factors encouraging, and sometimes forcing, men to work in the mines. These included the high wages offered by mining firms (relative to what could be earned in sharecropping) and gaining access to credit, although miners also made decisions about where to work based on factors such as the social scene and types of food available on particular concessions.

There is a need to move beyond oversimplified "push" and "pull" factors in our descriptions of African migrant workers and examine more closely the "in-between" logistics of such mobility. Although the work of some scholars has gone far to emphasize precolonial economic institutions, as well as modified versions of such, on the Gold Coast, researchers have missed important connections between rural and urban economic pursuits, respectively traditional and modern economic activities, which existed during the colonial period. This is where credit schemes have been receiving a growing amount of attention. The social securities provided by labor agents will also be examined.

The economic lives of African have been ignored or undermined in a number of economic studies in the past. Theories such as that of the "backward bending supply curve" gained prominence in the fog of habitual generalizations applied to indigenous economic activities and the ignorance that surrounded indigenous social and religious thoughts.[20] Followers of this theory posited that although wage rates acted as an incentive for many Africans, they could not be too high, because African workers had in mind specific earnings targets, which once met would prompt them to leave their jobs and return to idleness. This theory has been critiqued as presupposing that African workers were inherently lazy. It was this mindset that allowed certain colonial observers to defend low wages for African workers, for whom paid labor supposedly was a substitute for leisure activity. Missionaries on the Gold Coast also reiterated this idea when they described migrants from Liberia as "seeking a fortune wherewith to return home, purchase a wife or two, and settle down to a life of lordly ease."[21] In line with a number of recent studies demonstrating that wage rates in the Gold Coast Colony were generally high during the colonial period,[22] *Mediators, Contract Men, and Colonial Capital* shows that employers in the mines had to offer high wages to attract and keep laborers. Most contemporary foreign employers did not bother to consider alternative explanations for why local laborers preferred to work in a flexible manner or remain in agriculture entirely. Yet, more recently Raymond Dumett has described a number of obstacles stood in the way of an expansion of the wage labor market in a "classic" way, that is, the mobilization of local laborers for permanent work. Among other things, low labor density in the Wassa region, local laborers' acute awareness of the danger of this kind of work, and a social stigma against underground mining played roles. Instead of conceiving of the transition to wage labor as strictly an exchange of labor resources from family-based subsistence activities (during the precolonial era) to hired labor (during the colonial era), this study attempts to take into account the ways in which preexisting, precolonial socioeconomic conditions in West Africa informed the growing wage-labor market during the colonial period, though in particular, as concerns former slaves, a neat transition is not always easily empirically drawn.[23]

Colonial Commerce and African Agency

In the context of growing numbers of wage-earning Africans in colonial mines, many scholars were also concerned with the brutality of capitalism. In consequence, many of them also designated the colonial mine as a centerpiece of social revolution. Usually using Southern Africa as an example, they demonstrated the extent to which African workers were subjected to the whims of capital, citing as evidence the introduction of pass laws, the totalitarianism of mining compounds and oscillatory migration. In parallel, they showed how workers' resistance to these conditions arose. This narrative has also been applied to research on the early colonial Wassa mines.[24] Although studies of this kind can be lauded for shining a light on the precarities faced by the global working class, they have also tended to erase, distort, or exaggerate the agency of African workers to fit this framework. More recent scholarship has challenged the image of the omnipotent employer in the mines, with an eye toward greater details in the behaviors of African mining men, as seen in the debate surrounding "cheap labor theory."[25] This study revisits some of this material with a microhistorical approach to provide new insights into the behaviors of African workers.

In addition, this study follows in the footsteps of a number of recent studies that present gold rushes during the colonial period in Africa as not necessarily European-led. In the state of Wassa, colonial infrastructure also helped to revive traditional forms of gold mining. And most local Akan laborers still preferred to pursue traditional mining in combination with a rural lifestyle during the first gold rush of 1879–85. In addition, African merchants and traders from the Gold Coast played an essential role in promoting the gold-mining area and conducting land transfers on behalf of mining companies. Indeed, these agents founded some of the first mining firms on the London Stock Exchange.[26] The uneven yet rapid spread of capitalism in the southwest Gold Coast created additional openings for indigenous actors as well, namely the African middlemen who aided in the mobilization of wage laborers. For the vast majority of individuals attracted to or lured, tricked, or coerced into wage work in nineteenth- and twentieth-century in Africa and large segments of the globe, an intermediary was likely involved. Therefore, *Mediators, Contract Men, and Colonial Capital* also builds on the observation that the task of expanding the wage labor market for expatriate mining was neither "automatic nor inexorable."[27] Even more than in the major gold-mining centers of South Africa, the indirect recruitment of African contract men was crucial to large-scale production (even though only a minority of African workers opted to sign agreements for steady and extended work). Yet little information is available about the dynamics of bringing these several thousand largely male African workers to the mines each year on the individual level.[28] This study therefore demonstrates a fuller appreciation for the roles played by a range of African mediators in the expansion of the West African wage-labor market serving the Wassa mines. Although literature on these and similar figures in South Asia is quite rich, there have been fewer micro-level accounts of their activities on the African continent.

According to Jane Burbank and Frederick Cooper, a framework of analysis concerning intermediaries translates into the study of "people pushing and tugging on relationships with those above and below them, changing but only sometimes breaking the lines of authority."[29] These engagements, therefore, moved beyond a simple dichotomy of resistance and collaboration. This notion of "contested collaboration" is particularly useful, as it provides a more nuanced view of the colonial encounter. Several case studies have already demonstrated the great potential of this approach for a "new" colonial history In *Intermediaries, Interpreters, and Clerks: African Employees in the Making of Colonial Africa*, Benjamin N. Lawrance, Emily Lynn Osborn, and Richard Roberts show that African intermediaries "held positions that bestowed little official authority, but in practice the occupants of these positions functioned, somewhat paradoxically, as the hidden linchpins of colonial rule."[30] Similar figures also emerged in the context of colonial commerce to help manage larger African populations on behalf of Europeans.

Where African intermediaries have featured in studies on colonial commerce in West Africa, they have generally been depicted principally as agents of foreign financial interests who facilitated the economic exploitation of other Africans. Alternatively, they have been conflated with village headmen. More needs to be said about those labor agents on the margins of society who were neither fully anchored in the mining hierarchy nor held a firm place in the political structures of indigenous society.

The narrow discussion of the notions of authority and honor relevant to labor agents has only been matched by the underplaying of "struggle factors" in the balance of power. Struggles and negotiations occurred both between African intermediaries and African workers and between African intermediaries and European employers, as in the image conjured by Burbank and Cooper, though granted, when looked at over a broad span of time, the authority of African intermediaries did generally decrease during the colonial period. The experiences of African labor agents and the larger population of African workers, the agents' relationship to mine managers and the colonial state, provides an interesting lens through which to understand how colonial domination was manufactured—and also how African actors managed to shape this process after their own needs and desires. The actions, demands, and conflicts of African workers are explored.

Global Labor History

The question of where the socioeconomic histories of Africa and Asia converge with that of the "global North" is currently being debated more heatedly than ever before. Scholars have proposed that there ought to be a more concentrated effort to open up the field of social history in the global North, which had left out over half of the world's population, thereby obscuring many "similarities, differences and interactions between regions and continents."[31] As a result, scrupulous comparisons are now increasingly being made in order to get a closer understanding of patterns of capitalist intensification in different parts of the world. In labor histories of both "the West" and "the Rest," scholars had a longstanding tendency to bundle the spread of capitalism and "free" wage labor. As a result, a spotlight was constantly aimed at the (male) "free" wage worker. (And as mentioned earlier, colonial mines were frequently presented as a forcing ground of class consciousness.) Yet this new field of study advocates that social and economic historians broaden their gaze to include all workers, including sharecroppers, the unpaid (e.g., family or household labor), and the coerced (e.g., slaves or indentured servants).[32] The worker (as well as what constitutes work) is now being conceived outside of a limited, unilineal ideal. For as Gervase Clarence-Smith observed in reference to the modes-of-production debate, although the handling of models has stimulated some interesting research questions, it is also important not to be blinded by the dogmatism of such lines of inquiry: "This in turn provides a channel for the real to re-enter the rational."[33] Dumett's earlier research on the nineteenth-century Wassa gold mines also moved away from the application of classic Marxist theory when discussing the transformation of the labor force, since the fluctuating and seasonal nature of the mining labor force did not fit into the theoretical framework of class formation.[34]

Global labor history considers Africa a fundamental part of a broader discussion of social and economic development and capitalist intensification, without diminishing the importance of the its local contexts.[35] The shift toward what has been termed "hybrid Marxism" has also triggered renewed interest among historians to investigate the history of industrial labor relations in Africa with loosened theoretical restraints[36]—this study being a case in point.

In accordance with this field of inquiry as I understand it, *Mediators, Contract Men, and Colonial Capital* upholds the distinction of the *free* wage laborer, in the classic Marxist sense, and the *wage laborer*, as in someone earning a wage temporarily but not necessarily dependent on it for survival. A wage laborer in this sense could also be an *independent laborer* during other periods of the year. Nevertheless, this study does not reserve the title of *worker* for the permanently urbanized. A contract laborer with a six- to twelve-month term of service is also a *worker*, as far as this book is concerned.

The ramifications of indentureship also play into this topic. This study highlights the expansion of the wage-labor market in the western protectorate by way of voluntary (or choice-based) debt bondage; the incorporation of modified forms of traditional relationships of dependency into modern commerce. It does not discount African workers as workers simply because processes of choice-based indebtedness brought them into the wage labor market. Polly Hill, in particular, has warned against the standardization of a "colonial attitude to debt" in studies on African economics, which she states "was as moral as that of missionaries to adultery and polygyny."[37]

Indigenous economics and the endogenous factors shaping colonial capital in Africa have been studied from a variety of perspectives. And new source material for the gold-mining sector in the western region of the Gold Coast has allowed for yet another perspective on this period presented here, wherein the demands of industry had to adapt to the realities of economic and social conditions on the West African coast in order for a capitalist transformation to occur.

Sources

The frenzied gold rush that took place in Tarkwa in the late nineteenth and early twentieth centuries produced a broad range of documentation from Ghana, the United Kingdom, and other parts of the globe. Yet collecting the appropriate sources to reconstruct this narrative has not come easily. The boom-and-bust nature of imperial gold mining, combined with the reluctant involvement of the Gold Coast administration in Wassa, meant that much of the written materials that would be expected in other corners of the British Empire during this period were scarce or never preserved in the first place. The Ashanti Gold Fields Corporation, established in Obuase to the north of Wassa, left an extensive archive, including the Cade Papers, whereas most Wassa gold mines simply left behind slim piles of liquidation papers at Kew Gardens. Educated African entrepreneurs did not settle in Wassa, and in fact the administration actively sought to curb their business activities. Missionaries were late to settle in the area, with the first group being the Methodist Missionary Society in 1910. Furthermore, what sources have survived hardly make any mention of African labor, reflecting how mining entrepreneurs, colonial officials, and visitors ranked their importance and overall contribution to the mining sector. These silences proved to be a motivating challenge to this labor-centered study of the Wassa mines.

In the debates that took place around the viability of large-scale gold mining in Wassa, whether at the level of the British Parliament or within the colony itself, and in the surveys that were conducted as a result, British officials spoke of the African labor force only in an abstract manner. There were, for instance, data describing the demographic makeup and population size of African workers in the mines. Reports also discussed the range of coexisting forms of labor relations on local mining concessions. The private papers of Frederick William Hugh Migeod, the chief officer of the government transport department, who was stationed in the western region starting in 1903, diverged from formal administrative papers in ways that were clearly beneficial to this study.

I have to admit that I was taken aback by the depth of information about African miners available in those records held at the Cambridge University Library, a short trip from Oxford, my research hub at the time. Furthermore, they scarcely have been analyzed. While one other study on mining in southwest Ghana made use of F. W. H. Migeod's papers, as did a monograph about the elite class of colonial servants governing British West Africa, neither study had paid much attention to the details of labor organization. Few other documents contributed so much to my knowledge of indirect recruitment by indigenous intermediaries and the place of labor agents in the many-layered mining hierarchy. Another key contribution to this study was a collection of letters between African miners and family members back in their rural home villages, some of which are printed in their entirety in chapter 2. Although these letters are discussed in the context of workers' motivations first and foremost, they provide precious information about gender, familial, and financial relations in the villages connected to Wassa's migration networks.

Court records proved of greatest relevance to this study. Proceedings from the District Commissioners court in Tarkwa (ADM 27) are held at the Public Records and Archives Administration Department in Accra, Ghana, where I stayed for several months gathering data. I am not aware that this material has been used in other scholarship thus far. This was likely the case because the handwritten trial recordings are not easily legible. A few pages had been permanently lost due to pests' damaging the paper. Nevertheless, it was well worth the effort of going through each one carefully. The documents detail labor disputes severe enough to be heard in British courts starting in the 1870s and covering several decades thereafter. They were invaluable sources for discerning the nature of contestations and collaborations between companies and labor agents. The same goes for the agents and the members of their gangs. Records illuminated aspects of the credit relationships, which are elsewhere only discussed in broad and general terms. These documents were also useful when it came to distinguishing between differently motivated labor contractors in Wassa during the period under investigation, both individuals and agencies. Although only a few of the records actually discuss those marginal agents, who were tied neither to a particular company nor to any government institution, I can count myself lucky to have located at least some information regarding the business activities of these entrepreneurs.

Additional interesting material on the gold mines of the 1950s and 1960s, labor, and the political economy of the Gold Coast (and then

Ghana), not included in the final version of this book, was kept at the regional archives in Sekondi. I traveled to the archive, which at the time was located in a tiny office with two desks in a business building in the western region. Records here were fewer and not so well kept or organized as those in Accra. Still, they would be useful to anyone pursuing a study of mining labor in the era of decolonization.

The travel diaries of mostly low-ranking white miners or European visitors who recounted their time in Wassa proved to be an illuminating addition to the collection, and not least for the photographic materials they contained. Compared with those in official records, their descriptions were less restrained, and more vivid and detailed. Different individuals wrote about production processes (also from the layman's perspective) and the activities of African employees on different concessions. Indeed, it was in travel diaries that I encountered most of the details about the work of African women in the concessions. Reconstructing the histories of African women is not only a personal priority of mine, but in the spirit of global labor history this book would not have been complete without reference to women workers, their collective action, and the manner in which they adapted their skills to the demands of technological change. (A prominent female labor contractor is also mentioned in the Migeod papers.) All of these exciting and unexpected insights have mitigated any frustrations felt and the many dead ends encountered along the way in the creation of this book.

Structure

The story of labor mobilization for the Wassa mines during the first and second gold booms is told in four chapters. Chapter 1 provides a historical overview for the rest of the thematically organized chapters. At the center of this chapter is the theme of change: It demonstrates the transformations that occurred in managerial professionalization, capitalization, and the technology of mining over the course of the last quarter of the nineteenth and roughly the first quarter of the twentieth century. Professionalization, for instance, entailed an influx of prominent prospectors and promoters, many of whom were transplants from South Africa. Their perspective on the labor question was more cohesive, albeit narrow-minded, than that of mining entrepreneurs during the nineteenth century, who held a multitude of clashing opinions on the labor issue. Chapter 2 initiates the second major theme of this book, innovative continuity: It begins by introducing actual practices of labor recruitment in the nineteenth century. Before a detailed discussion about the origin, function, and pragmatism of the form of indirect recruitment practiced mostly by Liberian contract laborers, coexisting forms of mining labor in Wassa are introduced, including casual, "coolie," and tributary labor.

Chapter 3 examines the second gold boom to show that technological and managerial improvements had but little impact on the way migrant men were brought to the mines. In fact, the turn of the twentieth century marked a period in which recruitment from Liberia became too expensive, and mining companies were obligated to break ties with labor recruiters from that country, men with whom they had built long-standing relationships. Yet out of this situation arose business opportunities for a number of independent labor contractors who, unlike the Liberian labor agents, used their own capital to recruit workers from other parts of West Africa. These recruiters on the margins of any intentioned indirect recruitment system often hired others to command the gangs they were investing in. The ever-present threat of labor desertion was exacerbated through the practice of poaching by labor agents with no firms ties to a particular mining firm.

Chapter 4 introduces another intermediary body for the recruitment of mining labor, the government transport department. Beginning in 1903, officials re-created the indigenous recruitment and supervision system within this colonial agency in order to subcontract men to the mines. Nevertheless, with the proposal of a Concession Labor Ordinance that same year, the chief officer of the bureau anticipated stricter labor regulations, including pass laws, for individual laborers in the mines in the future. Ensuing debates help to bring out the tensions at various levels of the colonial state on the issue of disciplining mining labor and, related to that challenge, the future of economic development.

Chapter 5 examines the transformation of corporate indigenous recruitment as the mines began to recruit through chiefs in the Northern Territories in large number. This chapter ends with the later reinstitutionalization of the system in its more market-driven form.

1

Prospectors, Politicians, and the Question of "Progress"

The First and Second Gold Booms in Wassa

As was the case in other parts of the world, the discovery of gold in Wassa in the 1870s stimulated the development of infrastructure for the exploitation of this metal, as well as the implementation of governance mechanisms to keep in check a large wave of unruly immigrants. Yet the slow speed and limited capacity of such advancements in the region set West African mining apart. Many of the earliest entrepreneurs in Wassa during the first gold rush in the 1870s and 1880s were former political officials and local European traders who tried to keep up production in spite of rather costly and irregular means transportation, since their petitions for public investment and services for the mining centers largely fell on deaf ears.[1] Still, they hoped to make a considerable return from local gold deposits using alluvial mining techniques and light stamps that were generally transported to the interior either on the Ankobra river or on the heads of female porters. The construction of a government railway was slow moving. As a result, it took some twenty years for it to connect the area to the coast.[2] Still, this was an accomplishment that played a key role in attracting an influx of significant South African mining capital, which in turn drew in more experienced mining directors and working-class miners from Britain, South Africa, and Australia in the early twentieth century. Also instrumental to the making of the second gold boom, which was referred to as the "jungle boom" in the international press, was the installation of heavy machinery, which brought West African gold-mining capacity up to global technological standards starting in the 1890s.

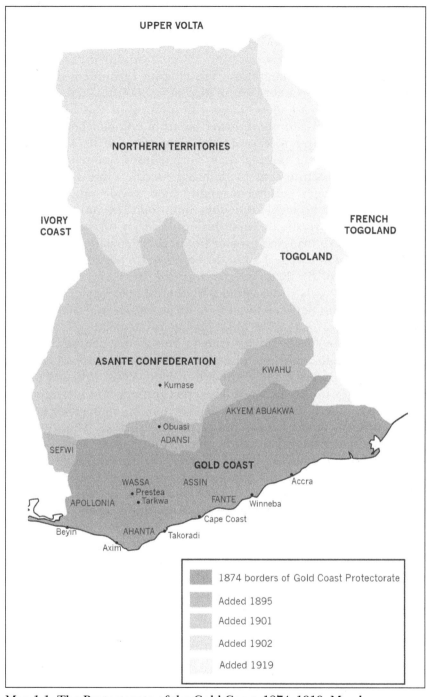

Map 1.1. The Protectorate of the Gold Coast, 1874–1919. Map by Talya Lubinsky.

Debates over the labor question show that if efficient transportation was slow to arrive, proper colonial administration lagged even further behind. Therefore, this chapter introduces the colorful group of entrepreneurs whose ideologies, politics, and cutthroat promotion helped to shape large-scale mining in this loosely governed British protectorate, located far from colonial headquarters in Accra. It attempts to capture the character and background of the most prominent mining men during both boom periods. In the nineteenth century, when lobbying for the mining sector was in its earliest stages, mining magnates held a wide range of opinions on how best to tackle the labor question. The outlook on this issue became more streamlined during the early twentieth century. However, besides bringing greater political organization and managerial skills, newly arrived men also imported a particular brand of racial division of the labor market into the West African gold-mining industry. This is demonstrated by the increasingly acrimonious daily interactions between lower-level white miners and African miners and the petitions of the Mine Managers' Association—a central organization for local mining firms in Wassa—for compounds and pass laws, though many would have considered themselves content with just the stricter implementation of labor laws. Still, any meaningful changes in the arena of labor recruitment and management hinged on closer collaboration between mine management and British officials in the colony and in the British Isles. That relationship remained fraught, if not neglectful, during the initial boom of 1879–85 and improved only slightly during the second boom of 1900–5. In spite of the gradual infrastructural and managerial improvements to the mines, and the outward projection of a more cohesive hard-line stance on the labor question after 1900, mine managers ultimately failed to build the political strength necessary to arouse serious interest in Accra or the Colonial Office. In the end, in the absence of any consistent or pointed intervention by the state for the purpose of locating and retaining a cheaper and more reliable labor supply, feelings of distrust continued to outweigh any cooperation between local managers. This chapter sets the scene for the remaining thematically organized chapters by illustrating how the contentious political and business climate in West African gold-mining production generated its own unique version of economic pragmatism, opening up new ways for indigenous actors to fulfill the mines' most basic demands for labor recruitment and retention.

The First Gold Boom

Timing was an important ingredient for the first gold boom in Wassa, because it "occupied a lull between the New Zealand and Colorado gold rushes of the 1860s and 1870s and the great stampede to the Rand in South Africa in the second half of the 1880s."[3] The year 1877 marked the beginning of prospecting in Wassa, which lay outside the boundaries of the Gold Coast Colony proper, and which had come under British protection just three years earlier. Not long after, a number of European entrepreneurs, as well as a few indigenous ones, began to acquire mining concessions for the purpose of either reselling them for a profit or establishing profitable mining firms. The Compagnie Minière de la Côte d'Or d'Afrique, or the African Gold Coast Mining Company, was first to register there, on April 10, 1877. Its founder was a thirty-three-year-old French trader named Marie-Joseph Bonnat. Bonnat, according to a fellow mining entrepreneur, was "a bright, cheery little Frenchman of great energy [with] some knowledge of the Fanti, or rather the Asante language, and perfect experience of the native character."[4] Coming from humble beginnings, he traveled to West Africa to make his fortune.[5] The tale of his journey to this corner of the world, which began as part of the Niger expedition of 1867, matched much of the drama that ensued during his stay there. According to Basel missionaries stationed on the Gold Coast, on leaving for the west coast of Africa, his ship was wrecked in the very first storm it encountered off the coast of France. Provided with another small vessel, he and the remaining passengers still managed to reach their destination. "Here [Bonnat] separated from his countrymen and began to trade, with the assistance of his mulatto servants, going further inland until he at length reached Ho."[6] A mere two years after his arrival he was appointed manager of the Basel Mission trading post there. Therefore, it was a marked change in his path when, in 1869, during the third Anglo-Asante conflict, Bonnat and a few Swiss missionaries were taken to the Asante kingdom as captives. This particular event, though likely traumatic, eventually brought him into intimate contact with the Asante people and their king and knowledge of the Wassa area. His treatment in Asante was not much different from that of other European hostages. Bonnat was often subjected to what was perceived as the king's arbitrary and capricious behavior. "At times denuded of every stitch of clothing, he was compelled to make mud bricks and at night was often tied to a tree so that he could neither sit or lie down."[7] Still, it was said that on other occasions "he feasted with the king and was shown all his treasures of gold, which were enormous."[8] According to rumor, vital details about the gold mining region came from the king himself. It was supposedly during one of these more amicable encounters with the Asantehene Kofi Kakari that Bonnat learned the origin of the gold bountifully adorning Kakari and other Asante royals. More important, during this period the Frenchman began to develop the types of personal connections that later facilitated business around Wassa, which he swore to visit on his release. Freed from captivity in 1873, Bonnat became one of very first Europeans to prospect for gold in the area. A mere three years later, following a lecture at the Paris Geographical Society on the potential of the West African gold fields, he successfully attracted enough capital from a number of French and English backers to establish a company to explore for gold. Initially, and in spite of his limited scientific knowledge of mining, he founded the Société des Mines d'Or d'Afrique Occidentale, a syndicate devoted to the acquisition of concessions. In 1877, during another trip to Paris, Bonnat became a founding member of the Anglo-French African Gold Coast Company, which later took charge of what would become the long-standing and highly productive Tarquah and Abosso concessions, which extracted 37,338 ounces of gold to be processed and sent to Europe between 1889 and 1900.[9] Between 1892 and 1896 the company's annual gross revenue averaged £10,940.[10] Bonnat's unlikely origin, as a former petty trader from France who began his career by purchasing "a quantity of cotton in exchange for cloth and powder"[11] in order to conduct trade along the coast and who ended up becoming one of the leaders of the mechanized gold-mining business in Wassa, would be replicated to varying degrees during the first gold rush.

A couple of months after Bonnat established his firm, a number of companies were founded, predominantly with British capital, in the mining towns of Tarkwa, Tamsu, and Abosso. Several successful concessions came to be managed by the Effuenta Gold Mining Company (1879), the Wassa Gold Coast Mining Company (1882), the Abbontiakoon Mines (1880), the Tamsu Gold Mining Company, and the Swanzy Estates (1881). Yet in a fashion typical for the contemporary gold-mining industry, investments into West African mining were highly speculative. It has been estimated that British investors annually paid upward of £9 million sterling into foreign mining companies during the 1880s, and this sum steadily climbed by the 1890s.[12] This considerable sum was nevertheless measured, since the mining companies were generally expected to prosper for a few years at most. More telling in the West African case is that the bulk of this money was pooled from a large number of small private investors, rather than from any major institutions, such as banks.[13] Interested parties were willing to take on high-risks for potentially high returns. During the first gold boom of 1879–85, James Irvine of Liverpool, Francis Swanzy of London, and the former explorers Richard Francis Burton and Verney Lovett Cameron were some of the prominent promoters of gold mining in Wassa who were willing to take that gamble.

Promoters' Perspectives on Labor

During the 1860s, James Irvine belonged to a class of general merchants acting along the West African coast. His main business interest was palm oil, a primary export of this area at the time. The slump in oil prices in the 1870s, however, pushed Irvine toward mining development. A "self-proclaimed pietistic Christian who took great pride in church membership, charitable public service, and circumspect business dealing, he exemplified the contradictions in European entrepreneurship in West Africa during the mid-Victorian period of informal empire and the 'three Cs' (Christianity, commerce, and civilization)."[14] On the one hand, he advocated for the social, economic, and political development of West Africa. On the other hand, he followed a cold and calculating business approach, which led him to spread misleading romanticizations of local production for the sake of inflating the prices of stock for his own personal gain, as well as that of his stockholders. In what was likely also a blatant attempt to further bolster stock prices in West African mining, he invested in the companies of several of his peers in Wassa. Yet his lack of a cohesive strategy in running individual concessions was perhaps revealed by the fact that he ended up backing concessions led by mining men, who among themselves could not have had more divergent opinions on the question of labor recruitment for the Wassa gold mines.

The successful Cinnamon Bippo (1889) was one example of such a venture. It was a joint enterprise between Irvine and Frank Swanzy, a fellow merchant with large West African assets. The Swanzy family had established a commercial presence on the Gold Coast starting in the late eighteenth century. Frank and Andrew Swanzy were also former palm oil traders, who formed mines as subsidiaries of their mercantile firm F&A Swanzy, rumored to be the oldest and largest European trading firm on the coast.[15] The firm first tried developing a quartz vein outside the town of Abosso, but without great success. This initial experiment, however, "led eventually to their working the continuation of the banket deposit at Abosso." Their success mining with this particular formation of ore prompted the creation of the prominent Wassa Gold Coast Mining Company,[16] which took over the Cinnamon Bippo in 1895.[17] A senior partner of the firm, Frederick J. Crocker oversaw work on this concession, where managers and engineers became notorious for welcoming the employment of local Akan casual and tributary laborers in staggering numbers, as will be shown in the following chapter.

At the same time, Irvine went into business with the former explorers Richard Francis Burton and Verney Lovett Cameron. In 1881, Irvine commissioned Burton, at the time the acting consul at Trieste for the Colonial Office, and Cameron, an avid traveler, to inspect and report on the Wassa gold fields. They arrived at Axim on January 25, 1882. Behind Irvine's solicitation was the calculation that, given their rank and celebrity, these two prominent men would contribute to the international promotion of West African mining. And according to plan, immediately on returning to Europe, Burton preached far and wide about the promise of the Wassa gold fields at a series of public events. His bold predictions were also published on the opinion pages of a leading trade journal. In one he pronounced, "I take courage by observing that the Gold Coast, which threatens to oust California from her present prominence, is deemed worthy of Yankee jealousy."[18] He went on to assert the following: "Most of the Gold Coast 'bush' is second-growth, far more easily cleared than that of the Californian foot hills when they were attacked by the 'Argonauts of '49.' The Yuba ridge of the Sierra Nevada was as densely wooded as most parts of the Gold Coast, and the further we go north we shall find fewer trees, more grass and a greater quantity of gold."[19] Speaking of West African mines in reference to other prominent gold fields around the globe, with an added tinge of exaggeration in favor of the former, was a habitual practice in contemporary mining promotion. Through yet another media channel, much of Burton and Cameron's journey to the Gold Coast was documented and published in the two volumes of *To the Gold Coast for Gold* (1883), in which they also repeatedly speak of mining conditions in West Africa in highly optimistic terms. Furthermore, in an exhibition of Burton and Cameron's personal affinity for Irvine, the book has the following dedication: "To our excellent friend James Irvine (of Liverpool, F.R.G.S., F.S.A. etc.) we inscribe these pages as token of our appreciation and admiration for his courage and energy in opening and working the golden lands of Western Africa."[20] Burton eventually sat on the board of Irvine's Guinea Coast Mining Company (1881) and even invested some of his own capital into local alluvial mining, including the Akankoo mine, though there is no evidence that he ever made any profit from his investments. Throughout his tenure as a mining entrepreneur, Burton was never shy about expressing his disdain for the hiring practices of Swanzy's Wassa Gold Coast Mining Company. He also frequently and insolently expressed his misgivings about the productive potential of African labor. Indeed, Burton became a leading advocate for the importation of coolie labor on a large scale.

African investors and promoters also collaborated with James Irvine. As it turned out, they could occasionally be the greatest champions of economic intervention by the colonial state. These were men with a radical modernist vision of progress, who believed that not only the Gold Coast but also a larger West African economy would reap huge benefits from mining development.[21] Some of the most famous African mining entrepreneurs, such as Joseph Dawson, James Africanus Horton, and Paul Dahse, were predominantly involved in the leasing and selling of properties in the area. In contrast to the second gold boom, however, during the earliest stages of mining development, Africans were also managers of productive concessions, as in the case of the Abbontiakoon mines (1880). And, as with European entrepreneurs, in Africans' endeavors promotion, management, and investment activities overlapped as well.

Perhaps the most avid African promoter was the Liberian-born Ferdinand Fitzgerald, the editor in chief of the London-based *African Times* newspaper. He was able to popularize his vision of West Africa's economic future before a wide and politically powerful black and white British and African audience. In numerous lengthy opinion pieces, Fitzgerald framed mining development as an opportune and important project that British officials were basically obligated to facilitate at the very least. He made the argument that supplying the mines in the western protectorate with the tools they needed in order to advance would not cost Britain a single penny in the long run, assuring his readers that it "would be mere temporary aid from a rich parent to a poor child."[22] He could heighten his already emotional treatment of the subject matter by instilling his requests with a sense of humanitarian obligation. For instance, in an attempt to portray Britain as practically retarding the economic potential of the region, in one of his editorial pieces, titled "The West Africa Gold Fields," Fitzgerald declared, "The garments of the boy of ten years will not adequately clothe the man of thirty."[23] He left this image to ripen in the minds of an audience basking in the glow of Britain's antislavery campaign—an audience that was now bent on realizing further social change in Africa. Yet none of these calls for British assistance contradicted the fact that Fitzgerald was also a staunch advocate of African self-determination.[24] A prominent member of the Fante Confederation, a political alliance of "nationalist-minded" African entrepreneurs and chiefs,[25] Fitzgerald had been an early proponent of British governance over West Africa during a stage in which the Dutch still maintained a presence in the colony. He backed Britain because, among other reasons, he saw it as playing a critical role in the military defeat of the Asante kingdom. In the 1860s, this particular political concern was directly linked to economic interests. In contrast to the British, the Dutch had a history of supporting the Asante, despite their causing frequent disruptions of trade between the interior and the coast. In a confidential letter to the Colonial Office, Fitzgerald declared his allegiance to the queen by stating that the British favored "the promotion of trade and commerce, the development of material resources, the maintenance of friendly relations with the interior, and the general education of the native populations; matters on which the future welfare, prosperity, and advancement of this town, and the interior connected with them, must necessarily depend."[26] In the 1870s, Fitzgerald continued to push a large chunk of the responsibility for the economic affairs of West Africa onto the shoulders of British public- and private-sector actors. At the same time, his goals were becoming more concrete. By now he had moved away from grand concepts such as "progress" and "civilization" in his writings to focus on specific items requiring immediate attention by the Crown in its newly formed colonial state; the Wassa gold fields, now in the international spotlight, gave him an ideal platform to express his vision.

Not long after Wassa became a protectorate of the Gold Coast Colony in 1874, Fitzgerald composed a lengthy list of demands for the colonial government in connection with the development of the mining sector, the two leading items being land privatization and a more sophisticated transport system. For instance, as early as 1875, immediately following the conflict with Asante and right before the legal emancipation of slaves had been enacted, Fitzgerald called for the construction of government roads and the clearance of waterways connecting the mines to the coastal region.[27] According to him, "The work of reconstruction must now be diligently carried on; and most influential in this work will be the formation of solid, permanent highways, for facilitating intercourse between the coast and the interiors, for cheapening the transit of produce and goods; and thus promoting the development of agriculture and those other rich sources of wealth with which a beneficent Creator has so richly endowed those inter-tropical Gold Coast countries."[28] On the questions of land ownership and labor, in Fitzgerald's various public and private correspondences he raised the subject of more transparent property rights with regularity. He predicted that the "infection [of land privatization] will spread among the natives. Kings and chiefs will see what can be done by improvement on their lands. A class of proprietors, with defined, transmissible, or hereditary titles, will strive to improve that to which he has an immutable title, and which he may sell, or give, or bequeath to others."[29] Furthermore, he believed that "giving a value and security of possession to land, will certainly attract capital,—capital will give beneficial labour employment."[30] In Fitzgerald's vision, hired labor would be the norm. In clear contrast to both Burton and the managers of the Wassa Gold Coast Mining Company, he vigorously supported the idea of disciplining indigenous labor, conditioned on changes in land rights. Until this task had been realized, Fitzgerald foresaw a region that would sadly remain a "mere swaddles baby, in comparison with the productive and commercial giant, as which it will be known hereafter."[31] Nevertheless, this great destiny would be far from fruition until "Great Britain shall awake to the consciousness of those vast resources, the key to which has been placed in her hands."[32] At such a point, he argued, "active life will displace stagnation":

> The people and the country will for the first time live. A genuine spirit of progress will be infused; and Christian civilization will, it may be hoped and expected, not only spread over and embrace the whole of the British territories (for such they must become for any good whatever be accomplished) but pour its vivifying rays far into those interior lands whose people are now continually looking and pressing towards that sea-coast, whence alone they can obtain the few products of foreign industry which they covet.[33]

Speaking as a gentleman of West African origins himself, the Liberian-born Fitzgerald trusted that the economic awakening of the Gold Coast Colony and its protectorates would have a spillover effect on the rest of West Africa.

Fitzgerald argued that the British Isles stood to benefit from West African economic advances. Mechanized gold mining in Wassa, as explained in his publication, would simultaneously advance the British cause of introducing the universal gold standard at a stage in which there was still much anxiety over its irregular supply. His friend James Africanus Beale Horton, who was a frequent contributor to the *African Times* in the 1880s, had the following to say on that subject: "Not only are the interests of England more deeply involved in this question of an adequate gold supply than those of any other civilized community, but it is pre-eminently her duty, as the great inventor and propagator of the gold currency theory and practice, to stimulate that supply by every legitimate means in her power."[34] Could West Africa possibly be the key to a more stable level of exploitation and global trade? In Horton's opinion, the rates of global consumption of gold and the very spread of capitalism all but mandated a new gold rush: "It is not only that the gold currencies need all, and more than all, the gold that is now yearly produced, but that the increasing luxury of advancing civilization is a glutton of insatiable appetite, capable of eating up another Australia, and still reiterating the horse-leech cry, 'Give! Give!'"[35] He was certain that if, "as is generally admitted, there be evidence of a gradual exhaustion in the gold fields of Australia and California, at the same time that the world's wants of the precious metal are extending and increasing, new fields must be energetically worked."[36] Under these circumstances, Horton declared, "we cannot but deplore, as a grievous error, the indifference, the absence of all effort—nay, more, the obstructiveness [*sic*]—of the British Government as regards the development of the undoubtedly rich gold fields of West Africa."[37] Ferdinand, the outspoken newspaper editor, would eventually sit on the boards of directors of several mining firms in Wassa.

All of the mining entrepreneurs mentioned above forecasted a bright future for West African gold mining. However, the path leading to that future remained murky at best, especially without sustained investment and administrative services from the colonial government. In the 1870s, mining companies continued to try to appeal to colonial and imperial interests. Initially they acted as individual entities but by the 1880s increasingly collaborated in pressure groups with other business interests from Manchester, Liverpool, and London.

The Political Economy of Mechanized Mining during the First Gold Boom

During the 1870s, the Wassa mines were already sending petitions to the state for long-term capital investments in infrastructure and governance capacities. The stakes, according to them, involved the likelihood of peace and order. Reliable transportation and the regular implementation of labor laws were other key issues they believed would affect the overall viability of large-scale mining—not to mention sanitation, liquor control, communication infrastructure to connect the coast and interior, the conduct of a mineralogical survey to aid mining firms in their pursuits, and the imposition of a new understanding of mineral land rights that favored fixed boundaries and private property for the area. In the 1870s, all of these ideas, ranging from highly ambitious to more pragmatic, were pitched to the Colonial Office and the colonial administration in a rather chaotic fashion, predominantly by individual mining firms. By the 1880s, more formal lobbies were established.[38]

The growing presence of prospective miners to the southwest made it impossible for colonial officials in Accra to completely ignore what was happening in and around the mines. Therefore, 1882 marked the appointment of the first commissioner, the Irish-born William Andrew "Tim" Cuscaden, to Tarkwa District. Though some observers may have interpreted Cuscaden's appointment as a victory for mine managers, who had requested a greater administrative presence, his actual impact on the area was miniscule, as he held no magisterial powers.[39] Indeed, Cuscaden was appointed "with vaguely defined terms of reference and little power in administering the law."[40] He was primarily on a mission to assess the actual likelihood of finding gold in large and payable amounts and to weigh in on the potential burden that the expansion of this industry would place on the administration. The commissioner's correspondence on colonial economic policy was sent to Accra in the form of numerous reports whose precise influence is impossible to measure. Nonetheless, they remain interesting because of the way in which they underline a number of the colonial administration's preexisting concerns about mining activities. As did Cuscaden, other officials questioned what the broader social and economic implications of the gold rush would be for Africans at all levels of society. For example, would traditional rule be able to withstand such change? They also pondered how to regulate companies and ensure that they contributed financially to the colony. In the end, the commissioner's overall assessment was a negative one, solidifying the ongoing practice of laissez-faire economics and strengthening the colonial state's disinclination to react to mine managers' requests for improved transportation in a speedy manner.

Following Cuscaden's first report on November 22, 1882, which was circulated in a confidential dispatch between Lieutenant Governor W. Brandford Griffith and Colonial Secretary John Wodehouse, Griffith made the damning prediction that "not a single mine in or about Tarquah will pay."[41] During his two-year inspection tour around the mining centers, Andrew Cuscaden (with the assistance of the temporary commissioner Henry Higgins) drew up the extensive, four-part *Report on the Gold Mines*.[42] His three primary conclusions, as outlined in the text, were that (1) mechanized mining was above all self-serving to individual prospectors and therefore unlikely to bring greater prosperity to the colony; (2) land grabbing in Wassa would cause political disruption; and (3) large-scale industry would be harmful to indigenous commerce, which had thus far carried the colonial economy. During his investigation, Cuscaden gained firsthand experience of the ugly side of an industry that was often highly dependent on unsubstantiated promotion for profit. He reported to officials in Accra that many companies seemed more interested in manipulating shareholders in London than they actually were in constructing a profitable business that would bring wealth to the region. A mining engineer from London working at the Gold Coast Mining Company at the time of Cuscaden's inspections shared some damning information with him to underline this point. The engineer claimed he had been offered one hundred pounds to produce a favorable report on local operations by "a gentleman who is at the head and tail of most of the mines in Tarkwa district."[43] Most likely this was James Irvine himself. Cuscaden further indicated that the frequent praise that the *African Times* had given to another one of Irvine's concessions, the Effuenta Gold Mining Company, was entirely ludicrous, because when two tons of quartz were purchased from the mine and shipped to England to be assayed, it "hardly showed a particle of gold."[44] These endorsements, as he saw it, had less to do with any real productive outcome on the concession than they did with Irvine's affiliation with the editor Ferdinand Fitzgerald, who was also serving as the secretary of the company at the time.[45] Commissioner Andrew Cuscaden alerted officials to the fact that "the statements made in England by people in connection with the mines are calculated to mislead the public."[46] Floating companies and later projecting profits on these speculative ventures were ordinary practices of many of these early mining prospectors.[47]

These unscrupulous entrepreneurs pushed Cuscaden to virtually disregard any imaginary future profits of the mining sector and instead concentrate on the problem of the potential financial burden that the industry could place on the administration, whose resources were already stretched thin. The political situation also remained volatile. Not only did he perceive many of the new mining entrepreneurs as selfish and greedy but he also judged them to be quite reckless when it came to planning. It was no secret that the West African climate could be unforgiving, nor that a great number of foreign workers did not adapt well to local conditions. As a result, the commissioner anticipated groups of European laborers and entrepreneurs arriving in the colony, falling sick soon after, and then feeling entitled to paid passage back to Britain financed by the colonial administration.[48] Cuscaden himself depicted an environment where on "both sides are stagnant swamps, the exhalations from which quickly affect the constitution of the European."[49] High mortality rates among white mine workers were already gravely diminishing levels of productivity on the various concessions in the area. It therefore became an imperative of the colonial state that a population rush from Europe, North America, and Australia be avoided. The potential health and transportation costs associated with the mining entrepreneurs, who were pushing the frontier of permanent European settlement into the rural interior, received attention because the embryonic state was simultaneously dedicating considerable manpower and resources to the military campaign against the Asante kingdom. At the same time, the consequences of potential pushback from the Asante were not illuminated in Cuscaden's arguments, because he was primarily concerned with the ill effects of a rising concession boom that coastal businessmen were orchestrating around the mining villages. In essence, he, like many other colonial administrators of his time, worried about an erosion of the system of traditional authority, on which the colonial state depended so heavily. Apinto in Tarkwa district was one example of an area supposedly ravished by land grabbing. Although his assessment of sudden rapid change was likely, at least partly, informed by the stereotype of stagnant African economic relations, the commissioner relayed that there was "hardly a square mile in the province of Apinto that has not been rented."[50] He lamented, "This indiscriminate granting of concessions will lead to serious complications in future, unless rules and regulation are established for the better guidance both of the Europeans and native landowners."[51] First, contestations were growing among chiefs and kings, who often did not have precise records of their land boundaries. Second, land grabbing had given birth to a new and powerful profession, brokering land leases. The traveling commissioner spoke of powerful, educated African businessmen, whose authority he captured with the title of "bush-magistrates." They were known to secure the loyalties and the land of local authorities strictly in order to resell or lease the land to third parties. The rising influence of this elite group of

African entrepreneurs in Wassa was especially worrisome because they had been operating entirely independently, overseeing all facets of the negotiations between local kings and chiefs and foreign businessmen. In a logical next step Cuscaden foresaw them holding significant power over what he described as foolish "traditional" authorities. Emphasizing just how easily they were deceived, he described the main local chief Ashanta, the blind king of Tarkwa, as "(nearly) blind in a physiological sense, too."[52]

The African Joseph Dawson of Wassa was one of the most infamous sales agents. Before serving as a political agent for French and British companies, the Methodist-mission-educated Dawson was an administrator of the Gold Coast Colony.[53] According to the civil commissioner, by the 1880s Dawson exercised[54] "the greatest authority over the chiefs and people in the district of Apinto, and indeed I may say throughout the whole of Wassaw."[55] He and other African intermediaries were accused of helping European mine managers dupe local chiefs in exchange for a slice of the profit. Certainly there is evidence to suggest that as Dawson's power in the leasing of concessions to foreign companies grew, so did his unscrupulous ambitions to promote the mining industry. For example, during the fourth conflict of the war between the British and the Asante, the manager of the Effuenta mines accused Dawson of deliberately withholding news of an imminent Asante attack and also of bribing local chiefs to do so as well, in order to keep production running. Demonstrations of power of this kind led Commissioner Cuscaden to believe that although the kings and chiefs claimed to rule the various stretches of land around Wassa, "they are completely in the hands of one or other of the European managers."[56] On a very basic level, according to the administrator, such unprincipled commercial activity needed to be contained because it directly interfered with the legitimacy of the system of indirect rule. Once these businessmen managed to cloud a chief's judgment with expensive gifts and embellished promises, colonial officials would have to question the latter's loyalty and affiliations to the state.

Toward the end of his report on the "bush-magistrates," Cuscaden claimed to have found an instrument to freeze the activities of these affluent African brokers: "I am glad to be able to report that I have succeeded in putting a stop to these careers. A circular issued by the late governor, Mr. Ussher, pretty clearly showed these persons what they might expect if they were detected."[57] Yet in spite of the legal measures taken in 1870s and 1880s, individuals such as Dawson, Paul Dahse, James Africanus Beale Horton, and Ferdinand Fitzgerald continued to exert a strong influence on the political and economic life of the region.

Another argument reinforcing colonial nonintervention in Wassa was the potential devastating impact that changing property rights could have on indigenous commerce. As far as Commissioner Andrew Cuscaden was concerned, independent miners "for a long time to come will be the real prospector[s] of the wealth of the soil of this country" and therefore should not be driven out of the area.[58] Indigenous small-scale mining had continued during the colonial era and even flourished with the advent of mechanized mining, partly thanks to improved roads to and from Tarkwa. Therefore, if privatization was inevitable, Cuscaden advocated in favor of securing small-scale African miners' continued access to land for digging gold, as well as for farming: "Right should be given [to] him [the traditional miner] to search, dig, and prospect wherever he liked outside the leased land, and when he is successful, a certain number of fathoms square should be granted [to] him free of royalty, as his property, from which he could not be driven so long as he kept working on it. Should, however, he give up work for any length of time, say one year, he should lose all right and privileges."[59] Indeed, the administrator suggested that the government develop privatization rights, but first and foremost for traditional miners, in order to further stimulate growth in that segment of the market.[60] He expected small-scale indigenous mining, which had been productive for centuries up to that point, to outperform large-scale mining for the foreseeable future. This was a reasonable conclusion, because independent mining by small groups continued to attract most Akan workers living in rural areas, even after the establishment of mechanized mines because, amoungst other things, it was more compatible with seasonal farming responsibilities and could supplement income after a particularly bad harvest.[61]

Although this book cannot capture each and every whim of officials in Britain or Accra, its more nuanced (less certain and less catastrophic) take on the work of the British Empire for the period under investigation remains intentional. Although Commissioner Andrew Cuscaden's reports in no uncertain terms supported the notion that the mining sector would not become an engine for economic development in the colony, there is no way to conclusively assess their influence on future economic colonial policy. Perhaps, then, the reports are more intriguing for what they reveal about the process of gathering and sharing information across the protectorate, and for showing that serious inquiries about West African mining were conducted in the first place. On that same note, it is also important to discuss how the mere structural dynamics of the governorship and the Colonial Office mitigated against any serious identification with the mines. In addition to actively shunning excessive financial or manpower expenditures for the sake of intensive economic development in West Africa, the governor and the Colonial Office faced unintentional hurdles as well. Certainly, state-led development had no primacy here to begin with, in contrast to other colonial territories in the British imperial system such as Ceylon, Jamaica, and South Africa. With that being said, frequent staff changes and rotations within the Colonial Office and the colonial state further stunted changes in West African policy. As Dumett asserts: "Even in case of routine approvals for municipal work projects, positive declarations issued under one administration were often shelved and had to be reaffirmed innumerable times by successors before implementation could be guaranteed."[62] In consequence, the larger narrative concerning the place of the Wassa mines in the colonial and imperial economic imagination in the 1870 and 1880s needs to be considered not only in light of the indifference that officials evinced toward the managers' loudest protestations. As a final point, it is important not to misinterpret the politics and pragmatic function of the colonial state in Wassa as being anticapitalistic. There were clear limits to its interest in protecting African laborers, just as administrators were reluctant to incorporate them into the institutional framework that the state provided: "It [the colonial state] failed utterly to investigate miners' living conditions or suggest administrative procedures that would benefit either the European companies or African workers in this period."[63] The colonial state expressed an overall lack of interest in overseeing industrial relations in the colony, identifying with *neither* the mining companies nor the African workers in their employ. This stance changed only gradually. For instance, by the 1890s, following a new wave of changes in the staff of the Colonial Office and the administration, some of the many proposals for mine management began to gain serious traction.

The Making of the "Jungle Boom"

The 1890s marked a turning point in the practice of and policy commitment to West African gold mining. Even though Britain's interest in its West African colonies continued to be primarily political in nature, an era of expansionist developmental imperialism was emerging, encouraged by staff changes in the Colonial Office, especially the election of

Lord Salisbury's Unionist coalition administration in 1895, and the colonial state, especially the start of the second administration of Governor W. Brandford Griffith in 1885. The newfound willingness to support private entrepreneurs at these two levels of the British Empire resulted in a more coherent outlook on the issue of economic progress. And the new colonial secretary, Joseph Chamberlain, for one, was determined to break the trend of colonial self-sufficiency. His overall strategy was to dedicate public funds from the British imperial government to lifting the colonies to a new stage of social and economic development. Among a number of actions taken, Chamberlain coordinated capital investments into infrastructural projects and scientific knowledge for the sake of supporting foreign and indigenous commerce.[64] Chamberlain's ambitions happened largely to align with those of the incumbent governor of the Gold Coast, W. Brandford Griffith. The turn of the twentieth century marked the moment when the colonial state finally gained a foothold in the region. As the internal structures and bureaucratic functions of the administration became more formal and specialized, it began to follow a more straightforward colonial agenda of territorial expansion and, most relevant here, the support of capital. Griffith was an enthusiast of progressive development policies and also happened to maintain a favorable outlook on mechanized mining. Following a visit to the Wassa gold fields in 1889, the governor expressed how impressed he was with the new managers in charge. He further declared that the area was "rich in gold" and it was "merely a matter of the necessary time and scientific application for that gold to pay well for extraction."[65] Under his direction, a variety of programs and projects were realized to grow the economy, with the government railway system likely having the greatest transformative impact on a variety of economic sectors.

The construction of the government railway in the early twentieth century opened up a new phase of colonial commerce by connecting remote rural markets in the colony and its protectorates to ports and peoples on the coast. Tracks ran between the port city of Sekondi and the rural mining towns in the southwest, reaching Tarkwa in 1901, Obuase in 1902, Kumase in 1903, and Prestea in 1911. Railway transportation brought significant relief from many of the uncertainties and costs involved in head porterage and transportation by boat. Since endemic sleeping sickness prevented the use of any beasts of burden, the employment of gangs of carriers had been the most widely practiced means of long-distance transportation that mines in the forest zones had used until this point. It mattered tremendously to productivity levels and company expenditures that heavy machinery could now be more effectively transported from the coast, as could staff and goods for trade and consumption. Another action by the government that advanced mining development was the British defeat of the Asante Empire in 1901, a critical moment for local trade and production. Although mechanized gold mining had not come to a standstill during the final conflict between the Asante and Britain, production was in constant peril. The Asante had attacked mining camps on numerous occasions in the past. As a result, any rumors of imminent attacks were often enough to convince laborers to desert their workplaces. For instance, the educated African Paulus Dahse, who managed the Effuenta Gold Mines Company in the 1880s, once wrote to District Commissioner Reginald E. Firminger of Axim to enquire directly about the advances of the Asante army in Wassa, for his entire Kru staff had recently abandoned the camps after receiving news to that effect.[66] This lingering unease among managers and staff about the vagaries of the conflict dissipated with the conclusion of the violence. Political stabilization encouraged many previously reluctant investors to reconsider their involvement in the gold business.

More fortuitous factors also contributed to the recovery of the mining sector. For one, the gold boom of 1900–1905 occurred when the political disruption of the South African War impeded the Witwatersrand gold rush, which had been ongoing since the second half of the 1880s. Gold mines on the Rand closed because of the South African war of 1899–1902. European financiers were therefore already looking for new ways to exploit the precious metal for a profit when the suggestion that gold reefs in Wassa resembled those of the Transvaal gained popularity. Several credible sources, including scientists and colonial agents, and the less credible promoters repeatedly affirmed the assertion that the reefs in Wassa showed the formation of banket just as they were known to occur on the Rand and nowhere else, for that matter. Indeed, when several engineers from the Transvaal visited Tarkwa in 1900, they announced the reefs of quartz in the area to be of a similar, if not better, grade than deposits in South Africa.[67] All of a sudden Wassa was making a name for itself once again, revamped as "the Johannesburg of West Africa."[68]

A large degree of the promotion of West African mining hinged on this very title for the next decade.[69] Second, by the 1890s, the gold rush of Western Australia was also becoming increasingly economically volatile. The stock market in West Australian gold mining was proving to be quite unpredictable, scaring away many former and potential

Figure 1.1. The government railway to Kumase and Tarkwa, ca. 1900–1904. Basel Mission Archive QD-32.032.0117. Photo by Max Otto Schultze.

backers. In summary, under just slightly different circumstances a significant portion of the capital entering Wassa during the second gold boom would have been invested in more developed industries. As the price of gold rose on the world market, and neither the Rand nor West Australia showed much hope of recovery, many foreign investors looked to Wassa to bring stability to the market. As a result, by the end of 1900, mining development in the western protectorate was in full swing once again, as indicated in the list of active mines in the Gold Coast in 1904 (table 1.1). Suddenly Tarkwa was being mentioned in the same breath as Kimberley, Bulawayo, Asante, Klondike, Nome, Cripple Creek, Kalgoorlie, Broken Hill, Wyalong, Mambare, Croydon, the Towers, Palmer, Raub, Sumatra, and the Punjab—sites of other recent hopeful mineral rushes.

The sheer number of companies registered in Wassa at the turn of the twentieth century reflected this great anticipation. Besides a core of roughly forty mines that had survived the 1880s,[70] almost an additional two hundred were now staking claims. Over the entire boom period

Table 1.1. Yearly output of gold from 1886 to 1905 in the Southwest Gold Coast and Asante Region*

Year	Number of ounces obtained	Estimated value in £
1886	20,799.06	74,829
1887	22,546.80	81,168
1888	24,030.625	86,510
1889	28,666.80	103,200
1890	25,460.30	91,657
1891	24,475.60	88,112
1892	27,446.06	98,806
1893	21,972.06	79,099
1894	21.332.18	76,796
1895	25.415.94	91,497
1896	23,940.66	86,186
1897	23.554.74	84,797
1898	17,732.70	63,838
1899	14,249.87	51,300
1900	10,557.37	38, 007
1901	6,162.99	22,187
1902	26,911.15	96,880
1903	70,775.20	254,790
1904	104,460.279	378,480
1905	171,149.723	653,820

*The large increase in the export of gold starting in 1902 was due mainly to operations in the Asante region.
Source: PRO CO 98/14, Report on the Mining Industry for the Year 1905, enclosure in Gold Coast no. 175, April 9, 1906, p. 14.

from 1900 to 1905, approximately four hundred companies were registered in the towns of Tarkwa, Tamsu, and Abosso.[71] In just the twelve months between September 1900 and September 1901, over 170 companies were formed with a nominal capital of £24 million and a share capital of around £4 million.[72] By 1902, about seventy-one companies had even begun to conduct mining activities.[73] Many of the most prominent properties had changed hands coming into the possession of new and more experienced managers, including Cinnamon Bippo, Effuenta, Prestea, Tarkwa Main Reef, Wassau, Wassaw West, Adjah Bippo, Abosso, Tarkwa, and the Abbontiakoon Mines.[74] One observer praised:

> The class of man which they now send to the Gold Coast is the right one. The manager is a man who understands men and organisation as well as mining; the doctor—and a great deal in a mine in a tropical climate depends on the doctor—is a man who knows that there are other ways of keeping men fit besides the daily dose of quinine; the superior staff—the heads of the machinery, surveying, assaying, and other departments—is composed of men, each a technical expert in his own line; the foremen, engine-drivers, mechanics, and miners are, with the inevitable exceptions, sober, clean living, and hardworking men.[75]

This was a tangible change for the better, although the extent of local improvements should perhaps not be overstated. Others remained cautious in their praise. As late as 1903 a report from the *Economist* was adamant that the "West African Gold industry as yet can hardly be said to have emerged from the development stage, though according to some of the numerous mining companies' reports recently issued, development on a few of the properties is in a forward condition."[76]

A swelling population of young white men of varying experience, whose opportunities had dried up on the gold fields of Australia and South Africa, pushed the mining frontier in Wassa further forward, eventually outnumbering whites in Accra for the same period. Whereas 176 men of European descent were officially residing in Tarkwa in 1901, only half that number were in Accra, where 96 Europeans were stationed or living.[77] And the 1902 census for the Gold Coast Colony and its protectorates clearly attributed mining prospects to this growth. The document actually linked the increase in white men in the colony to two factors: the "great" development of the mining industry and the construction of the Sekondi–Tarkwa Railway.[78] In addition to a couple of skilled engineers and prospectors from the previous gold rush who were now going after jobs as mining consultants, a flood of European, American, and Australian miners and entrepreneurs, who had previously been part of the company networks, were now arriving at Wassa. A significant segment of these men maintained occupational and financial links to South Africa in particular.[79] With mine managers and foremen accustomed to being part of a transient work force on short contracts, recruiters for the mining firms simply had to make them aware of how financially rewarding just a few months of work in West Africa could be. And with most having previously worked on the Rand, they were not intimidated by the difficulties of "driving niggers," to use a contemporary phrase. The new mining men commanded a great deal of respect from colonial observers. Only on the rarest of occasions did one still come across "one of the old sort."[80]

If these newcomers consciously tried to distinguish themselves from the previous generation of foreign miners, their efforts certainly did not go unnoticed. Colonial commentators generally praised this new group of miners for having more of a long-term perspective on development. They were also said to be better educated and more efficient and economical than in the past. A visitor to the area around 1905 acknowledged how the "class of European miner in West Africa has improved out of all recognition in the last few years."[81] According to observers they added to the business in a wide array of aspects: "The result of this improvement in the staff is that money is well spent and not frittered away on ill-considered plans: the work is better done, and more of it, both by the European and native, for there is a better understanding between the two; and there is far less expensive wastage in men through ill-health."[82] In reality at least some degree of what was perceived as a "better understanding" between white and African miners was built on greater structural inequalities in professional terms for African workers.[83]

Ultimately, the dominant picture painted in the press was one of a rejuvenated and bustling town. Conquered by a whirlwind of human activity, the environment in Tarkwa, especially, was characterized by the roar of machines, as captured in the diary of one visitor.

> All that could be seen of the town from the station was a row of wooden, tin, and brick buildings, consisting of the Bank of British West Africa and various merchants' stores on the ground floor, with the living quarters of the occupants above them, surrounded by wide verandahs and covered with the usual whitewashed galvanised-iron roofs.
>
> To the west, on the other side of the railway, was a bare hill, dotted with blackened tree stumps, from which the Court House and

Figure 1.2. Managers and mineworkers at the Apinto mines in Wassa, ca. 1895–97. The National Archives UK, CO 1069-34-35.

two or three bungalows blinked at us in the dazzling sunlight. Several lonely-looking banana-trees were the only natural ornaments left on the whole hillside, for, as is generally the case with clearings out here, the work had been done too thoroughly. One would have thought that a few trees might have been left to please the eye and give a little shade.

North, south and east, the view was cut off by other hills on which were perched the buildings of Abbontiakoon, Tarkwa, and Effuenta. The tall smoking chimneys of the two former; the faint clanking of chains, rattle of stone, and buzz of machinery; and the bustling figures of numerous workmen, small in the distance, gave the scene a busy modern air that seemed out of place in the wild forest through which we had rolled in such leisurely fashion all the morning.[84]

The pace of change could have been swifter. Still, Tarkwa stood out from most of the colony and its protectorates as a model for economic "progress." Technological advances were quite astounding.

Management was sound. Professional foreign entrepreneurs, applying scientific knowledge, introduced a new method of deep-level mining to the area that required larger capital inputs in drilling, crushing, extraction, and separation processes.

South African Mining Networks and Technological Advancements

The incentive to discover new and profitable deposits in gold around the globe generally had the effect of animating transnational and transregional flows of capital, labor, and technology, as well as ideology and culture. In Wassa, these influences took on a more distinctive shape during the second gold boom. The frequent, short-term circulations of people around imperial gold booms and busts contributed to the leading mining men's lack of any in-depth (or any) experience of life and business in West Africa. They also gave no indication of taking any interest in the particularities of local socioeconomic circumstances when formulating plans for the future. The dominant vision was one of stark top-down change, to be achieved with assistance from the Colonial Office and colonial government officials. Yet as will be shown in the rest of this chapter, although they were successful at securing some support from the government, especially in the arena of transportation, more fundamental means of colonial assistance—such as greater legislative, executive, or judicial oversight in Wassa—remained elusive.

Percy Coventry Tarbutt

The most powerful new mining men in Wassa, the so-called jungle magnates, were connected to Cecil Rhodes's British South Africa Company. These "Randlords," who controlled much of the diamond and gold mining industries in southern Africa, also became invested in West African gold mining once the time came to find alternatives to Johannesburg mining. The Tarbutt Group, led by Percy Coventry Tarbutt, Edmund William Janson, and Edmund Davis, was the most prominent example of South African mining interest at work in Wassa. According to the *Financial Times*, this particular group of mines "led the van in the 'Jungle' Boom."[85] Speaking of Tarbutt especially, the article stated, "Those potentialities were foreseen by him before the big boom in W. Africa, and, being early in the field, with his friend and colleague, Mr.

Edmund Davis, he had become a Jungle magnate, with large and widely ramifying interests."[86] Before arriving in West Africa, the British-born Tarbutt had gained fame running companies all over the world. He was known to be a highly eloquent and persuasive promoter of gold mining in parts of Australia and North and South America. And he would later put these skills to use in support of mining developments in West Africa. Tarbutt initially moved to southern Africa in 1888; there he found employment as a consulting engineer for Cecil Rhodes's Consolidated Gold Fields of South Africa. This was where he became one of the early proponents of the novel technique of deep-level mining. At this moment, when deep-shaft mining was on the brink of becoming a widespread and influential practice in the industry worldwide, Tarbutt was pushing for its incorporation in South Africa as well as north of the Limpopo River. In spite of Rhodes's initial skepticism of this costly investment, the sinking of boreholes eventually paid off, yielding the huge quantities of gold lying several hundred feet below the outcrop of the main gold-bearing reefs on the Rand. By August 1892, less than a decade later, Tarbutt had become the director of Consolidated Gold Fields. He subsequently became a mining entrepreneur in his own right, founding a number of companies including the African Estates Agency, which purchased shares in other highly profitable deep-level mining companies. After returning to England with great wealth in 1895, Tarbutt went into the mining consulting business. He formed Tarbutt Sons and Janson together with Edmund Janson and Cecil Quentin, with whom he had run a civil engineering business, the Tarbutt Liquid Fuel Company, before leaving for South Africa.[87] The bond between Tarbutt and Janson was further cemented when the thirty-one-year-old Janson married Tarbutt's daughter in 1899. Though there were no obvious business collaborations between Tarbutt and Edmund Davis predating their time in Wassa, both men fostered close professional ties with Cecil Rhodes. The Australian-born Davis had also built a solid career in gold mining on the Rand. Arriving in South Africa around 1883, he initially made a name for himself in the copper-mining industry. In 1895, Davis and Rhodes cofounded the Northern Copper Company Limited, which merged with several smaller businesses to become the Rhodesia Copper Company in 1902. This company later established two of the most significant mining companies in the Copperbelt Province of southern Africa, Bwana Mkubwa and Roan Antelope. Davis earned the nicknames "Napoleon of Northern Rhodesia" and "Chrome King" due to his ownership of numerous chromium mines on three different continents.

Tarbutt, Davis, and Janson together founded the firm Percy Tarbutt and Co., which directed approximately one third of the active Wassa gold mines during the second gold boom. They took over several well-known and promising concessions including the Abbontiakoon, Apinto, Cinnamon Bippo, Adjah Bippo, Quaw Badoo, Effuenta, Fanti, the Wassa (Gold Coast), and Wassau West Mines, as listed in table 1.2.[88] According to a report in the *Economist*, Edmund Davis personally had a hand in no less than thirty-five companies in Wassa during the second gold boom, with a shared capital of 3,300,000 pounds.[89] It was stated of Tarbutt that his "capacity for work, his mastery of detail, and what may be called his generalship, were so remarkable that he held simultaneously directorships of no fewer than twenty four mining development, and investment cos. [companies] not all of which were African."[90] He was also chairman of three of these companies, including the British Gold Coast Company, Limited; the Mashonaland Agency, Limited; and the Village Reef Gold Mining Co. His uncompromising spirit was remembered as well: "As a director he was able in administration, with the advantage of practical skill in mining matters, and he was not the sort of man to be easily influenced by timid counsels or peevish protests when he had made up his mind for what he considered the best."[91] This inflexible characteristic would encourage the type of hard-line stance on labor that eventually brought him, and the mine managers whom he represented in the Mine Managers' Association, onto a collision course with officials in Britain and Accra.

Certain factors made colonial administrators weary of these new mining men still. Shareholder market manipulation continued to be an issue in the mining industry after 1900,[92] and it is quite evident that Tarbutt became rather ruthless in that respect. He and his partners opted to build their long-term outlook on mining in West Africa on an intensive promotion strategy. A 1901 report for the *African Review*, for example, illustrates one of his tactics of projecting the earning capability of the Rand onto the Wassa gold fields, using weak empirical evidence.[93] In the report, Tarbutt elaborated on the "enormous amount of old native workings along the outcrops of these banket reefs," framing them as a "first indications of the richness of this area."[94] According to him there were such workings over a twelve-mile area of the reef, attaining a depth of seventy feet, where the Africans had reached their limits but where machines could now take over. Although his statements up to this point certainly expose an interesting and logical, yet perhaps underexplored, relationship between the small-scale mines that Africans

Table 1.2. Percy Tarbutt's business activities in Wassa

Company	Issued capital in pounds sterling
British Gold Coast	100,000
Caida (Wassaw)	248,776
Managing Director	
Fanti Consolidated	412,764
DIRECTOR	
Abbontiakoon (Wassaw)	500,000
Abosso Deep	5,000
Banda Syndicate	10,600
Effuenta (Wassaw)	423,500
Fanti Mines	1,139,000
Gold Coast Investment	400,000
New Gold Coast Agency	549,997
Tarkwa Railway and Mines Syndicate	150,000
Wassau (Gold Coast)	246,300
West African Gold Trust	152,600
DEPUTY MANAGER	
Adjah Bippo	70,000
Cinnamon Bippo	92,000
TOTAL CAPITAL	4,510,537

Source: "Death of Mr. Percy Tarbutt," Financial Times, June 1, 1904, 4–5.

continued to work and the larger, mechanized mines, in his final conclusion, using rather tenuous deductive reasoning, he boasted, "This would indicate that the reefs were richer than those of the Rand, because there were no ancient workings at all along the whole of the banket series of the Rand."[95] At the same time, as already mentioned, there is little doubt that his firms also introduced a much higher level of production and management to their properties in Wassa. A short overview of one concession demonstrates this transformation.

The Abbontiakoon Mines after 1900

Located a few miles north of Tarkwa, the Gold Coast Mining Company's Abbontiakoon Mine was abandoned after the economic bust of the mid- to late 1880s. Originally under the direction of Ferdinand Fitzgerald and James A. B. Horton, on January 27, 1901, it was newly registered to the Davis, Tarbutt, Janson Group and floated with a nominal capital of £500,000 sterling.[96] Two hundred thousand shares were issued at one pound each.[97] A good amount of the profits from the shares was invested in newer and more efficient technology. Colonial commentators already regarded this property as being at the forefront of technological experimentation and advancement during the 1880s,[98] and this reputation largely remained intact during the twentieth century. When Decima Moore and Frederick Gordon Guggisberg toured the Abbontiakoon (Wassa) Mines during his assignment on a special survey of the Gold Coast Colony and Asante on behalf of the Colonial Office in 1905, they encountered a vibrant scene, which she described vividly in her travel diary:

> The top [hill on which Tarkwa stood] had been completely cleared of trees and presented a busy scene of noise and activity. The whir and clank of machinery from the big galvanised-iron workshops; the irregular tip-tapping of hammers and the steady seething-sound of saws from some new buildings around which a crowd of natives were busily swarming; the clattering of stone and earth as trucks, appearing suddenly out of the ground, shot piles of ore into wooden bins—all these sights and noises combined in proving that Abbontiakoon, anyway, was a live mine.[99]

Almost all aspects of mining had been updated during the previous two years. The government railway ran straight through the property "from end to end," allowing for easier delivery and transfer.[100] Moreover, the plant and machinery necessary for conducting boring operations had arrived early on.[101] On February 15, 1901, diamond drills produced by the Sullivan Machinery Company of Chicago arrived from Liverpool to test the richness of the various reefs. Six experienced American drill men accompanied them to the West African forest zone. Then in 1902, ore of a free milling character, which required the simplest form of treatment, was struck.[102]

By 1904, the main shaft of the mine, constructed at an incline of 35 degrees, was excavated to measure 787 feet, with 5,644 feet of levels that were driven into the reef along the seam of gold. Decima

Guggisberg's diary provides an account of the production process, starting from underground: "At intervals along the level, and at right angles to it, short galleries called 'rises' and 'winzes' were driven into the reef. The ground between the rises was blasted out with dynamite cartridges, care being taken to leave a sufficient number of pillars to support the roof."[103] Laborers hauled the ore up to the mouths of the shafts in small iron cages, called skips, on light rails. From here, it was gradually moved to the crushing houses. African laborers were in charge of carefully sweeping the pieces onto a slowly traveling foot-wide leather conveyer belt, which then made a fifty-yard path to the house and back along pulley rollers. Before chunks of the solid mass reached the building where they would be treated, other laborers, including women, inspected the various pieces of ore, checking to remove those containing no gold. "Inside the crushing house the ore was transferred to a sort of travelling staircase, which ran the lumps mechanically into the mouths of the 'crushers,' where the rattle of their fall was lost in a peculiar grinding noise." At this point the Krupp ball mill was activated. This pulverizing machine rapidly and without much effort worked its way through great amounts of ore. Made out of heavy steel and iron, the grinding mill reduced all remaining lumps into roughly walnut-sized chunks of ore and fed them into rotary driers, where they were rid of all moisture.[104]

The separation procedure also experienced improvements in its efficiency. Instead of using amalgam plates to recover the freed gold following pulverization, a chemical cyanide treatment process was applied. The fine, powderlike material that was the outcome of the pulverization process eventually landed on another conveyor belt, from which it was emptied into large, circular cyanide vats where chemicals aided in the recovery of a larger percentage of gold than had ever been possible.[105] In 1905, the treatment plant at the Abbontiakoon Block 1 concession treated 16,544 tons of ore with a diminished recovery of 8,721 ounces of gold before the end of the year, due to a fire ripping through the main shaft.[106] In 1906 it crushed gold up to the value of £132,681.[107]

Tarbutt and his staff made sure that the Abbontiakoon Mine was one of the first companies to break the dependency on firewood for fuel. Timber was ample in the forest areas surrounding the mines, but its supply was irregular as was the quality of the material. In addition, after several decades of mining, wood reserves were receding and had to be transported from farther and farther away from the mines. Wood fuel also consumed too much labor. Firewood had to be collected in a number of arduous steps. It was chopped and brought in on the heads of contract, but also on occasion piecework, laborers.[108] As table 1.3 demonstrates in 1904 Abbontiakoon the was one of the highest exporting mines around Tarkwa. In 1905, the installment of a nest of eight coal-fired boilers transformed the energy supply.[109] Coal, though no cheaper than firewood,[110] was especially convenient for those mines situated along the railway, because they were able to rely on regular deliveries from the coast.[111] Its introduction also freed up these laborers to follow other tasks on the property.

By the end of his career, Tarbutt was primarily associated with West African enterprises. In spite of his early death just years into the jungle boom in 1904, his impact on the trajectory of mining development in the region was undeniable. His admirers remembered him in all glory as "a pioneer of the movement for the development of West Africa's gold resources."[112] They placed hope in the industry's momentum which "though uneventful for the time being, is still fraught with great potentialities."[113] Tarbutt was also outspoken about the white mining force in Wassa.

Gerhard "Stocky" Stockfeld

The "South Africanization" of the Wassa gold mines took place at all but the lowest level of the mining hierarchy. In 1901, Tarbutt hired a Mr. Saltmarsh, who had previously worked on the Rand for the Consolidated Gold Fields, as chief engineer of the Abbontiakoon Mines.[114] The two men working under him, a Mr. Flowers and a Mr. Bradshaw, also had "considerable experience in South Africa."[115] The most notable of Tarbutt's staff was Gerhard "Stocky" Stockfeld, who became general manager of the Abbontiakoon Mines in 1904. Stockfeld, another Briton with significant experience in South African gold mining, eventually became a leader of West African mining development. He began his career in Western Australia, in the Kalgoorlie gold fields. After floating several companies there, Stockfeld accepted a contract placing him in southern Africa for what was reportedly a considerable salary. He moved to the region in 1895 as the leader of an Anglo-Australian prospecting party in Matabeleland. However, these plans were disrupted by the First Chimurenga—the uprising in Matabeleland and Mashonaland. Stockfeld even joined British troops to fight against Ndebele and Shona people for some time.[116] He remained in southern Africa for a few years, only returning to Australia in 1899 to serve as general manager of the Burraga Mine of the Lloyd Copper Company. In 1904, he took up managerial duties in Tarkwa for the Abbontiakoon Mines.[117]

This British mining man eventually became a celebrity in the Wassa gold fields, praised by his peers as the man "to whom is due the chief credit of having brought West African gold mining through troublous times to a successful issue."[118] A political official who held Stockfeld in the highest esteem asserted, "I fancy a good many 'Jungle' mines would have a different tale to tell if there were more men of Mr. Stockfeld's capability and energy in West Africa."[119] Other commentators insisted that he was "alone among the earlier managers, who showed real faith in the country; his name should not be forgotten."[120] According to popular accounts, no other mine manager matched him in optimism and persistence. Stockfeld's doggedness led him to employ some of the same promotional strategies as did his predecessors. He also overemphasized the similarities between Wassa and the Rand. For instance, in 1905 he insisted, "In most cases this habit may be called an injustice to the managers of the mines [on the Rand]; but I can nevertheless assure you that if you take the trouble to hunt up the records of the Rand at a period in its history which will be on a par with the age of the Tarkwa group, you will find that the comparison as regards costs will be in favor of the Tarkwa group of mines. . . . If our costs are at present higher than those of the Rand, our profits per ton are greater than most of the Rand companies."[121] Stockfeld also managed the Tarkwa and Abosso Mines,[122] which were under the operation of the Aqua Mining and Exploration, until June 1911. Apparently, his dedication to this company also encouraged the ingress of further South African capital into West African mining. Indeed, according to some colonial observers, "It is not too much to say that his success at the Taquah and Abosso was the chief reason for the South African houses taking an interest in the district."[123] He left a lasting impression as a dynamic manager and loyal promoter of the Wassa gold fields long after the end of the second gold boom in 1905. For although the Tarkwa and Abosso mines went into voluntary liquidation in April 1923,[124] years later, in the 1930s, Stockfeld was still publishing promotional pamphlets for the gold mining in Wassa. These included *The Gold Coast and Ashanti Gold Mining Industries: A New Rand*; *The Empire's Future Goldfield: A Second Rand: Gold Coast and Ashanti*; and *The North Ashanti Mining Company*.

The Tarkwa and Abosso Mines

The Tarkwa and Abosso mines were among the few well-financed, successful properties with no ties to the Tarbutt network. Having acquired two of Marie-Joseph Bonnat's prosperous concessions during the first gold boom,[125] Thomas F. Dalglish re-registered the company in 1899. Dalglish had previously worked on the Rand on behalf of the African Gold Coast Mining Company.[126] In 1901, Dalglish split the company, forming the Tarquah Mining and Exploration Company Limited and the Abosso property, a subsidiary. Dalglish ran the two concessions separately, with the Tarkwa concession having a considerable shareholding in the Abosso Mine. W. H. Rundall was mine manager until George Bailey took the reins in 1904. Stockfeld was general manager at the time. Though some milling had been conducted on a small scale on the Tarkwa and Abosso concessions from 1892 to 1898, after 1901 the mines were overhauled, and milling became a mainstay of production.

Located in the river valley approximately four miles south of Tarkwa, the Abosso concession was one of the big hopes for West Africa. In what was a highly speculative and sober move, the mine started with an issued share capital of £400,000. The Abosso concession boasted the deepest shaft in the colony in 1905. Of two shafts, the main incline shaft was 975 feet in depth with six levels, most of which were still being regularly expanded.[127] The level of production increased steadily with the erection of a cyanide plant and a thirty-stamp battery to replace the old ten-stamp battery that year.[128] As a result 26,843 ounces of gold, valued at £107,344, were crushed.[129] By 1907 the company was equipped with a new fifty-stamp crushing battery.[130] In addition to stamp milling and pebble milling, mine management also constructed a chemical treatment plant. A Cornish beam engine removed water from the mine at a rate of five thousand gallons per hour.

The Tarkwa concession underwent a tremendous amount of development, starting with a share capital of £349,079. Like the Abosso concession, which was located in the immediate vicinity of the government railway, the Tarkwa property was another easy choice for the installation of heavy machinery. In fact, the "Taquah mine was not only more favorably situated for work than the Abosso, but the reef being larger and more regular, and confined between walls, offered much greater facility for cheap work."[131] Underground labor was employed to remove the ore. Deep-level mining was developed at a similar pace at Tarkwa as on its sister mine, so that at the end of 1904, the main incline

Figure 1.3. African miner standing in front of a mine shaft in the Gold Coast Colony, ca. 1900–1904. Basel Mission Archive D-30.22.15. Photo by Max Otto Schultze.

shaft reached a depth of 750 feet.[132] Miners worked on 2,219 feet on three levels.[133] In the last four months of 1905, a light ten-stamp battery was introduced that crushed 1,605 tons of ore to make 1,215 ounces of gold.[134] Property managers constructed a cyanide plant that same year. It worked by a suction gas producer for a more economical use of fuel. The cyanide plant treated 2,360 tons of ore, recovering 673 ounces of gold.

The outlook of many of these companies was excellent. Investors back in Britain reacted positively to the infiltration of this new South African mining network. And, as shown in table 1.3, the scale of operations picked up dramatically over just a few years, producing a growing level of gold exports—from about 6,163 ounces in 1901 to over four times as much in 1902, to about 71,775 ounces in 1903, and about 104,460 ounces in 1904. Nonetheless, deep-level mining required a cheaper, more efficient and better disciplined labor force, not necessarily a shrinking one. In spite of certain dramatic historical changes in finance, technology, and management between nineteenth- and twentieth-century mining, in terms of the mines' relationship to the British

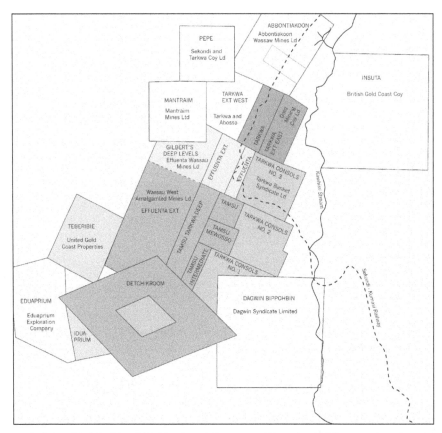

Map 1.2. The Tarkwa district showing concessions in 1906, snapshot from a survey conducted by the Gold Coast Government under the direction of Major A. F. Watherston. Map by Talya Lubinsky.

administration much also stayed the same. The novel demands of production, as well as political turbulences in certain parts of the fluid West Africa labor market, which had been supplying large numbers of laborers to the Wassa gold mines, would make recruitment an arena of even greater concern than before.

Table 1.3. Active mining companies in the Gold Coast colony* in 1904

Wassau (Gold Coast) Mining Company (Adja Bippo, Cinnamon Bippo concessions), situated along Tarkwa banket reef
Bibiani Gold Fields Ltd.
Broomassie Ltd., situated in Prestea and Bogosu district
West African Hinterland Consolidated Ltd. (Maymahalla concession)
Abbontiakoon Block I, situated along Tarkwa banket reef
Abosso Gold Mining Company, situated along Tarkwa banket reef
Himan Concessions Ltd. (North Bogusu concession), situated in Prestea and Bogosu district
Prestea Mines Ltd., situated in Prestea and Bogosu district

Other companies working their concessions:

Anfargah Gold Mining Company	Appankran Concessions Ltd.
Appantoo Mines Ltd.	Ashsnti Propriety G. Mines
Asiakwa Hydraulicking and Mining Company	Fanti Mines Ltd.
Attasi Mines Limited	
Axim Mines Ltd.	Gold Coast Amalgamated
Bartie Corporation	Muntriam (Wassaw) Mines Ltd.
Bokitsi Mines Ltd.	Nanwa Gold Mines Ltd.
British Gold Coast Company	Sekondi & Tarkwa Company
Chida (Wassaw) Mines Ltd.	Tarkwa Banket Mining Syndicate
	Tano Bippo Gold Mining Company Ltd.
New Gold Coast Agency	Tarkwa & Abosso Gold Mining Company
Eduapriem	United Gold Coast Mining Company
Effuenta (Wassaw) Mines Ltd.	Wassaw West Amalgamated

Dredging companies:

The Ankobra (Tarkwa & Abosso) Dredging Company
West African Mining & Dredging Company
Pritchards Dredging Company
Birrim Valley Gold Mining & Dredging Company

*Not including the Asante region
Source: Report on the Mining Industry for the Year 1904, CO 98/14.

2

Labor Recruitment in the Nineteenth Century

The Place of Practicality

As we saw in chapter 1, rapid commercial expansion and the loose political relationship between the mines and the colonial administration promoted a special brand of economic pragmatism in West African gold mining. This chapter examines the consequences of these conditions through the lens of the labor market. Although the Gold Coast's emancipation proclamation of 1874 might have helped to bring some individuals into the employment of the mines who otherwise would have been forced to work for their masters, most former slaves did not seek out wage labor as a primary means of supporting themselves. And although the mines drew the large part of their labor force from southern Ghana during the first and second gold booms, many mine managers still viewed local labor as a problem that needed solving. Local Akan men and women tended to work on a casual and flexible basis, which in the eyes of the managers was too costly and inconsistent an option. On the other side of the labor spectrum stood one particular category of employee: the "agreement boy," or contract man. Most managers agreed that it was vital "to encourage the boys to work on contract."[1] As one manager in neighboring Asante explained, "We get more work done and its [sic] far cheaper [and] it gives better satisfaction to both parties."[2] Men working on a contract-basis were favored by the companies because they required less supervision and less immediate capital due to the extended pay system. Furthermore, over the course of time they achieved a level of skill and efficiency that was essential to underground mining. Migrant laborers made up the bulk of this group in the nineteenth century. They were overwhelmingly Liberian Kru[3] laborers, whom foreign employers often referred to as the "Irishmen of West Africa," due to their humble yet hard work in the burgeoning palm-oil trade, as well as their diligent service in the British Royal Navy's antislavery patrols. They became the mining sector's most valuable workers during its initial decades of existence for their ability "to handle machines, to stoke, and generally to show the other natives both how to work and to exercise care."[4] Nevertheless, mine managers did not always get what they wanted. And in spite of their craving for even more disciplined labor, contract labor still failed to eclipse other forms of employment by the twentieth century.

With the onset of mechanized mining after the 1870s, Wassa's reputation quickly grew both in the Gold Coast and in its protectorates, as well as in other parts of West Africa, making it easier to recruit from these different areas. Yet an awareness of the sheer number of jobs available was in itself not usually enough to pull in potential workers. Many factors still discouraged long-term employment here, including the hazards of underground mining, the foreign time discipline, poor housing, a lack of interest in undertaking a long residence in one of the busy mining centers, and the racial ideology and lack of cultural sensitivity that increasingly informed work relations, to name a few points. Thus, in this context of labor shortage emerged a most crucial figure in the recruitment process, namely the indigenous labor recruiter, who offered managers a faster and easier means of getting large numbers of men to the mines by way of temporary agreements. Different recruiters were driven by different considerations. Some aligned their goals with the companies,' hoping to build a lasting career working for Europeans. The future and authority of others, such as the chiefs and village headmen, were rooted in indigenous authority. Still others demanded little respect in either of these spheres but were simply out to make a profit along the way. Liberian labor agents, who were dominant during the first gold boom, belonged to the first category. It was a logistical challenge to move large numbers of men from that country to the Gold Coast. Therefore, to do so, the labor agents bolstered their own authority by showing off their formal written agreements with a given mining firm, which incidentally also allowed them to travel by ship to the port of Sekondi at the firm's costs. Additional capital used to provide the men with food, credit, and small *dashes* (or small gifts) also came for the mines. Outlining the process by which contract workers were brought to the mines and held there, and the ways in which recruits, recruiters, and employers in the mines sought to shape this endeavor, are the central concerns addressed in this chapter.

It would be misleading to use the introduction of West African contract laborers into Wassa as a point of departure for a discussion on labor recruitment for the mines. After the end of the transatlantic slave trade, most employers in large-scale businesses within the British Empire began employing indentured laborers from India or China. Only on occasion were African contract laborers mentioned in the same breath. During one of these rare instances, following his visit to Tarkwa in 1889, Governor W. Brandford Griffith concluded that "the natives of the country cannot, unfortunately, be relied upon to supply the necessary labor, which must come either from the Kroo country or from China."[5] The general wisdom among employers was that Asian workers had something to teach the African ones. Therefore, although this chapter intends to capture the diversity of the labor market in Wassa, in terms of its demographics and the forms of labor relations generated by foreign mining development, it also discusses these factors in terms of the perceived vertical relations between different types of laborers from the employers' perspective.

Casual Labor: Capital Accumulation and Collective Action without Dispossession

Casual forms of labor, consisting of an amalgamation of piecework and casual labor not secured through a contract, contributed to the expansion of colonial commerce to a great extent. Although casual laborers were rarely mentioned in Wassa gold mines records, they consistently outnumbered contract men for the period under investigation. These were local Akan people who were simultaneously invested in independent forms of work, including farming and small-scale mining, and who had no wish to adhere to the strict regimentation of contract work. As for women in particular, managers would not have permitted them to live in the mining villages.

Many Akans perceived underground mining, where most contract men were engaged, as a realm of physical danger and spiritual darkness. To many, this work was as degrading and dishonorable as that of slaves. Others simply preferred to spend the bulk of their time working independently. This was understandable, because the traditional mining industry had taken off in parallel to mechanized mining during the first gold boom and promised greater wealth than could be accumulated through regular wage work at the time. Also, casual work did not interfere with farming obligations during the rainy season but provided

Figure 2.1. African surface miners, one receiving ore and the other deciding which stones are gold-bearing before throwing the rest out, next to a European overseer in the Gold Coast Colony, ca. 1900–1904. Basel Mission Archive D-30.22.14. Photo by Max Otto Schultze.

a path for the accumulation of extra-subsistence income, for instance when their crops failed. For all of these reasons and more, Akans chose to engage in temporary jobs in surface mining, which involved traditional forms of digging and extraction either above ground or in shallow, open pits.

Official statistics clearly reflect the overwhelming presence of casual employees: A survey taken in the colony in 1898 found that 2,032 surface workers, who predominantly worked on a casual basis, worked alongside 881 underground workers, who mostly worked on agreement.[6] A similar ratio is observed during some of the most productive years of the second gold boom. For instance, in 1903, 4,053 surface workers were employed on the gold mines, compared with 2,727 underground laborers. By 1904, the ratio was 10,941 surface men to 6,103 men working underground. Further, according to the survey "Labour in the Mines in the Gold Coast and Ashanti in 1904," Fanti, Asante, and Nzeman workers—groups living around major gold mining centers in protectorates of the Gold Coast—respectively made up 38.4 percent, 16.22 percent, and 1.42 percent of the mines' labor force overall, pointing to an extensive population of local workers in Wassa. Yet perhaps it has to be emphasized that these divisions (migrant-contract-underground mining vs. local-casual-surface mining) were not entirely clear cut. According to the Gold Coast Departmental report from 1903, the Yoruba from Nigeria, the Ewe from Togo and the eastern protectorate of the Gold Coast Colony, the Mende and Timmani from Sierra Leone, and the Vai people from Liberia all preferred to work above ground alongside local workers.

A number of concessions in Wassa also engaged tributary workers practicing traditional mining techniques.[7] Doing so allowed the properties to keep up production when other laborers could not be found or during times when cash flow was low. Tributary labor in its commercial form emerged as a means of monetizing land rights during a time when most lands were not under private ownership.[8] During this particular era and in the context of land abundance, land-owning chieftaincies tended to encourage the occupation of their lands by foreign and local producers, whether for digging minerals or for farming. In return, the chiefs demanded a form of tribute as a token payment in recognition of their continuing and rightful ownership of the land. In cases where they were dealing with their own subjects, they usually asked for one third of the produce or profit, whereas foreigners or migrants generally had to return two thirds of their earnings. Therefore, the fact that mining companies in Wassa in the 1870s adapted this system, expanding the types of economic relationships that they were willing to maintain with African workers, to some extent reflected a coming to terms with local economic conditions. The system implemented by the mines was more or less identical to what had existed before: They invited groups of independent laborers to work on their properties. The laborers did not receive a wage but instead retained between one half and two thirds of the gold-bearing ore they had dug up on a given day.

Nevertheless, as already mentioned, not all mine managers were as open to the incorporation of existing patterns of labor. Some preferred to impose what they contended were progressive labor relations. In the last quarter of the nineteenth century, there was vocal opposition to the use of the tributary system by local mining firms, especially by the entrepreneurs Richard Francis Burton and Verney Lovett Cameron. They lamented that the "country-people, Fantis, Accra-men, Apollonians [Nzemans] of [Beyin], and others, will work, and are well acquainted with gold-working; but they work in their own way."[9] These British adventurers appeared to have been dismayed by the practice of using tributary men, which had become commonplace on concessions such as those owned by the Wassa (Gold Coast) Gold Mining Company. The two former explorers were among those entrepreneurs who refused to invite any independent laborers to work on their properties without proper supervision. Moreover, they spoke derogatorily of those employers who chose to do so. It had come to their attention that a few competing mines were supporting these groups materially by providing them with the new technologies to make their work most efficient. In their travel diaries, they recorded that the "French mines supply them [tributary laborers] with tools and [dynamite] powder, and, by way of pay and provisions, allow them to keep two-thirds of the produce."[10] They denied that this habit of lending tools to these groups could benefit companies that needed to generate greater output. They also did not care that independent, small-scale miners regularly returned to such engagements when given the opportunity to explore new technologies. Instead their distrust of tributary men, which largely rested on concerns over theft, deepened. Burton and Cameron claimed, "It is evident that such an arrangement will be highly profitable to the hands who will pick the eyes out of the mine, and who will secrete all the richest stuff, leaving the poorest to their employers."[11] Stealing gold would harm the companies' bottom line. And many managers believed that the theft of equipment such as dynamite was on the rise. Given the purported disadvantages that casual labor posed to large-scale mines, it was that much more alarming that one company decided to wholeheartedly embrace tributary men en masse in its labor organization. This was the Wassa (Gold Coast) Gold Mining Company, which ran the Adjah Bippo and Cinnamon Bippo concessions and belonged to the Swanzy Estates. The manager of these properties, Frederick J. Crocker, installed a number of individuals with extensive knowledge of the region and its people. There was, for example, the educated African William Edward Sam, a former trading agent of the Swanzy company. In spite of Sam's respectful position with the colonial state, Crocker enticed him into leading the new venture at Cinnamon Bippo together with his sons, Thomas Birch Freeman Sam and William Edward Sam, Jr. Thomas Birch Freeman had acquired scientific training in mining in London. Indeed, in 1882 Governor W. Brandford Griffith described him as "a young native gentleman, who has been educated at the Crystal Palace School of Engineering, and is, I should think, a most reliable man who had some formal training in engineering."[12] Together the Sams and the brothers Louis and David Gowan were depicted as "different from the other managers" in their style of management. The strategy of voluntarily hiring large numbers of tributary workers in particular was linked to them.

When Commissioner Andrew Cuscaden visited Adja Bippo that same year, he commented that "a system of tributing exists here by which the natives of the neighboring village Ebukroom Indjim are allowed to extract surface ore, half of which is supplied to the mine and half they crush and wash themselves."[13] The governor also noted that Crocker "also mines at Adja Bepow on a different plan from the others; instead of paying a fixed wage per diem to the miners, each shaft is leased to certain people who receive one third of the output as their payment."[14] Of the 230 African laborers on the payroll of Crocker's property in 1889, only thirty were listed as Liberian Kru men.[15] It is unlikely that engaging tributary men was quite as backward and damaging to properties in Wassa as Burton and Cameron contended. Nor was it simply an easy solution for lazy managers.[16] Some managers beyond the Swanzy properties fell back on tributary work in response to the specific habitual shortcomings of their companies, for example, in times of capital shortages or when machines were defunct. In 1889, when the crushing stamps at the African Gold Coast Company broke down, the manager reportedly handed over more than 600 tons of quartz ore to tributary laborers, allowing them to retain one third of the gold that they could acquire through the process of pulverization, washing, and separation.[17] However, it is also improbable that managers on the Adja Bippo and Cinnamon Bippo works were forced into this position for reasons of low capitalization. After the death of Andrew Swanzy, Francis Swanzy showed an increased yet still cautious interest in mining development.[18] Though the firm's local mining investment continued to be a low-priority venture geared predominantly toward supplementary income, managers on these properties would have had a higher availability of loans due to their linkages with their parent trading firm. Most convincing is the suggestion that their flexible, or compromising, stance on employment was mandated by a need to uphold good relations with local authorities as well as other local consumers in the southern Gold Coast, ultimately for the sake of promoting their primary business.[19]

If tributary laborers largely worked in small open pits, their counterparts doing other tasks in surface mining were the piecework laborers. They, for instance, collected and cut firewood, cleared bush, and moved loads across concessions, economic activities that easily could be married with seasonal farming, and independent small-scale mining. Also, given the low level of mechanization, there were plenty of labor-intensive jobs in the extraction and separation process before the 1890s. One of the earliest accounts of a casual laborer, or rather, the leader of a gang of casual laborers, is the life story of the labor agent Coffi Accra, an individual "employed occasionally at task work" at the Abosso Mines.[20] In 1883, Accra was recorded as managing a group of men on the property of the Abosso Gold Mining Company, or Compagnie des Mines d'Or Abosso.

Not much was known about him other than that he had been a successful traditional miner in the past. Accra had dug his first pits with the permission of the local chief, Enemil Kwao Kuma or his predecessor, in 1866, and before long he had become a reputable independent gold miner.[21] When the land he worked on was sold to the Abosso Mines, "certain mining rights were reserved to [him]" as well as to other traditional miners.[22] This arrangement seems to have worked smoothly for several years before the concession became fully productive in 1881. During that time, Accra was assigned "to work certain shafts on the property on tribute" in addition to occasional work as a porter.[23] To this purpose he constructed a small camp on the concession, wherein resided the eight to ten workers who were part of his gang. Mine management had little input over either the hiring or the supervision process that Accra conducted. Indeed, the manager, Arthur Bowden, admitted to neither really understanding the nature of the relationship between Accra and his men nor knowing how he got them to work in the first place. At one moment, Bowden referred to the laborers as "principally relatives" of Accra's, indicating that they maintained an intimate bond.[24] At another, he described Accra as the "headman" of the part of the village where these men lived, implying a hierarchical political relationship.[25] Yet Bowden's confusion was not entirely misplaced, for a degree of mystery surrounded the composition of all gangs of casual laborers. For that reason, it would be ingenuous to believe that more severe forms of coercion did not inform at least some of these relationships. Slavery remained widespread on the Gold Coast even after the legal end of slavery, and new forms of dependencies were easily embedded into structures of family labor. It would have been just as easy to bring pawns into modern commerce in this manner. For instance, it is not unthinkable that mine laborers who fell into debt with local moneylenders were relegated to improving their status in this manner.

Local Akan women also engaged in piecework, completing tasks that conformed to traditional gendered mining roles. They initially provided only their washing skills to the mines in situations where machines were defunct. Mine managers were eager to let them take the lead in all decisions related to panning for gold. According to one,

"This is peculiarly women's work, and some are well known to be better panners than others; they refuse to use salt-water, because, they say, it will not draw out the gold":

> They have nests of wooden platters for pans, the oldest and rudest of all mechanical appliances. The largest, two feet in diameter, are used for rough work in the usual way with a peculiar turn of the wrist. The smallest are stained black inside, to show the colour of gold; and the finer washings are carried home to be worked at leisure during the night.[26]

These traditional skills were not fixed, but changed over time, for example, with the introduction of new technology after the 1890s.

Not unlike other piece workers women received task-based wages and did not find themselves to be in an essentially vulnerable position by means of their casual form of employment. In addition, not only was their wage rate more than double the average daily income of an agricultural worker for the same period, but also the following arrogant statement by a mine manager shows that they had little trouble bargaining for better work conditions: "They [the women] rarely wash more than 40 lbs., or a maximum of 50 lbs., per diem; and they strike work if they do not make daily half a dollar (2s 3d) to two dollars."[27] The collective action of not just women, but women working on a casual basis in the 1880s, is historically significant because it undermines the argument of free labor being a prerequisite to such political actions; though "free" labor conditions certainly facilitated collective action. Thus, the statement that "the individual contract of employment was the sole ancestor of the modern collective dimension of the labor-management relationship" may not ring entirely true.[28] Recognizing these early struggles for adequate compensation and fair treatment helps us to further dismantle a teleological view of labor market developments.

At the same time, it is also important to acknowledge that the overall position of women in the mines did not go from strength to strength. Instead, as the number of casual laborers exploded during the second gold boom, women workers slowly vanished from the various concessions. According to census data in table 2.2, 221 women were working in the mining industry in 1898, but by 1903 the number had dropped to 7. However by 1904 not a single woman was recorded as working for a gold-mining company. During this same period the male population of surface laborers rose from 1,811 in 1898, to 4,046 in 1903, and to 10,941 in 1904.[29]

Table 2.1. Persons employed at Gold Coast gold mines, 1898

Underground	Aboveground		Total	Total
Males	Males	Females		
881	1,811	221	2,032	2,913

Source: Mines and Quarries: General Report for 1898. PP, part 4 in Cd.112 (1900).

Table 2.2. Labor in the mining industry of the Gold Coast and Asante, 1905

Year	Underground	Aboveground			Total
	Male	Male	Female	Total	
1903	2,727	4,046	7	4,053	6,799
1904	6,103	10,941	0	10,941	17,044

Source: Mines and Quarries: General Report and Statistics for 1904. PP 1906, part 2 in (Cd. 2734, 2745, 2911).

The Asian Labor Question in Wassa

A sizable trade in indentured Asian laborers increased after the end of the transatlantic slave trade and spread in the context of imperial expansion. It was not only the product of empire but also helped to build empire itself. For many African colonies, importing so-called coolies was key to economic growth. This trade out of Asia also stretched to the southwestern Gold Coast, where rapid economic expansion was underway, after the 1890s. Mine managers in Wassa discussed the importation of Chinese and Indian indentured laborers from other parts of the British Empire as an attractive prospect right from the very beginning of the mines' existence. Using their correspondence with officials, the following section examines the distinctive language that was used to describe Asian indentured laborers. What characteristics and stereotypes were attributed to them? How were they perceived, also in relation to African contract laborers in particular? Finally, why did their success in Wassa not materialize as predicted by so many powerful and experienced mining men?

Employers in the Wassa mines rarely conflated the categories of African and Asian indentured laborers, Kru laborers being no exception.

Even after decades of recruiting African contract men through indigenous labor agents, many employers still envisioned a future in which Asian indentured laborers, especially the "industrious" and "dependable" Chinese, would provide a lasting solution to the mining sector's lingering labor problems. Burton, who made no secret of his reluctant reliance on African labor, was one of the strongest advocates of this scheme, promoting it whenever the opportunity arose. He and other proponents framed the propaganda surrounding Asian indentured labor as a means of engaging disciplined, reliable men who were not averse to hard labor under harsh tropical conditions. He also contended these men, many of whom already had mining skills and experience, would agree to terms of service of one to two years or even longer—a great improvement on the sometimes twelve-, but generally six-month, contracts agreed to by Africans. It was probably also easy to imagine a great number of Chinese working in the Wassa mines because no such things as a white working class, whose position could be threatened by their presence, existed here. Ideologically speaking, from the perspective of mining entrepreneurs Asian indentured laborers were a far cry from the British working class but decidedly a step above the supposedly lazy African worker, who was "incapable of regular and continuous labour."[30] This caste-based view of the global working class, which gradually intensified in West Africa over the course the two gold rushes, ultimately meant that the real forces guiding the choices of African workers would continue to be undermined and ignored, the workers' behavior being deemed both innate and irrational.

Foreign employers all along the West African coast had been toying with the idea of bringing in Asians on a large scale during the nineteenth century. An isolated few had even taken action. Burton asserted, "The whole West Coast, Liberia included, is crying out for coolie labour, and the French of Senegal have affected the savvy importation; already their trains are beginning to run. We questioned everyone who could enlighten us, and we did not hear a dissenting voice, except from Government officials, whose predilection for *quieter non movere*, for letting sleeping dogs lie, for reporting things pleasantly, and for ignoring unpleasantness is but too well known."[31] In trying to shake up the pragmatic economic policy of the Gold Coast, this British entrepreneur loudly professed that "the rest of the West African Coast is a luxuriant waste for want of coolies."[32] He also addressed the mining sector in specific: "If we are to oust California," he explained, "we must mend our manner of labour."[33] Together with Cameron he declared, "We must come, sooner or later, and the sooner the better, to a regular coolie-immigration, East African, Indian, and Chinese."[34] Revealing aspects of their racial and ethnic prejudices, they imagined drawing "freely upon the labour-banks of Macáo, Bombay, and Zanzibar" where they would encounter groups of workers to obey the demands of industry. The "intelligent, thrifty, and industrious Chinese"[35] would pick up mining "with the utmost readiness"[36] as they had done elsewhere; the East Indian "will be well adapted for lighter work of the garden and the mines."[37] Last, as carriers and laborers, the "sturdy [Swahili] of the East African coast" would achieve [three times the work of Fantis and Apollonians [Nzemans]."[38]

There is some indication that mining entrepreneurs also believed in the potential of African workers to adjust their work ethics to some degree. With that in mind, they put forward that the Asian presence would help African workers transform from their state of backwardness and laziness to a more advanced stage of productivity. They made the following declaration: "The benefit of such an influx must not be measured merely by the additional work of a few thousand hands."[39] For Africans, being confronted with Asian indentured laborers would "create jealousy, competition, rivalry."[40] The mining entrepreneurs insisted, "It will teach by example—the only way of teaching Africans—that work is not ignoble, but that it is ignoble to earn a shilling and to live idle on three-pence a day till the pence are exhausted."[41] Once these workers' industriousness rubbed off on local African men over the course of time, perhaps the challenges associated with supplying labor to the gold mines would be solved once and for all.

Initial attempts to bring a large number of Asian indentured laborers to West Africa occurred as early as 1882.[42] And after the 1890s, changes in staff at the higher levels of the colonial administration prompted the first experiments to bring them to Wassa. In 1896, William Edward Maxwell, the newly appointed governor of the Gold Coast, envisioned a mining sector in which miners who were "more industrious and better instructed than the Gold Coast negro" conducted alluvial mining.[43] His vision was realized through multiple experiments in 1897, 1902, and again in 1914. Although the British public rarely reacted positively to such trials, on all three occasions about thirty Chinese laborers arrived in the Gold Coast to engage in an experiment that proved to be a total failure time and time again.[44]

Although European employers in the colonies were invested in keeping the indentured laborers healthy and productive, unpredictable climatic conditions could easily throw off these efforts. And as one missionary stationed in the Gold Coast concluded, the "climate seems to be as deadly to the Oriental as it is to the European."[45] In Wassa many of the indentured Chinese laborers succumbed to illness and disease, so others were unwilling to follow in their footsteps. Therefore, in spite of all the excitement built up around the proposal of their introduction to the Wassa mines, gradually mine managers were forced to accept the possibility that, "only the African himself will ever be able to make available for the world the wealth which his country is ready to yield."[46] For many mining entrepreneurs this scenario was a troubling one, and as a result plans to bring Asians to Wassa continued to resurface for decades to come. That being said, the Asian indentured labor scheme also had vocal critics. Kwabena Akurang-Parry has described the political backlash to what the local Gold Coast intelligentsia viewed as the ultimate denigration of character of the African worker.[47] In addition, adverse voices also arose from within the mining sector itself. Picking up on discussions that were occurring in South Africa at the same time, Louis Gowan of the Swanzy Estates proclaimed, "I think it will be ruinous to employ Chinese labour."[48] Activating another set of racist stereotypes, only seemingly in defense of local labor, he described Chinese laborers as "pig-tailed cut-throats" who would contribute nothing to the West African economy.[49]

Indentured Contract Labor from Liberia and the Question of Coercion

There is no clear date marking the arrival of the first Liberian mineworkers to Wassa. By the 1870s, Liberian contract men were already extensively interwoven into the fabric of foreign commerce in West Africa, encouraging a British missionary on the Gold Coast to describe them as "probably the finest specimens of manhood in West Africa."[50] They had sold their labor power all along the coast of West Coast starting in the early nineteenth century and were frequently associated with work aboard ships: "Every steamer trading at the ports takes on a crew of Khroo-boys to do the heavier work such as the handling of the cargo, while the ship is in tropical waters."[51] They were also indispensable to trade, with many trading houses employing them to roll the puncheons of palm oil and kernels from the bush to the factories to the beach. The Kru also played a key role in transportation: "In many of the towns they form the crews of the surfboats, which ply between the steamers."[52] Their long-standing interaction with foreign employers along the West African coast was also documented in the nicknames they used in work interactions, including "Half-dollar," "Cash," "German-fast-boat," "Newspaper," "Pea-soup," "Tom Peter," "Broken-bottle," and "Black-man-trouble," to mention a few.[53] What can be said with some degree of certainty is that the first Liberian contract laborers heading to Wassa around the 1870s and 1880s first arrived at the busy port of Sekondi, a hub of commercial activity. German steamers, either from the Woermann Line or owned by the trader August Humplmayr, passed through here. These vessels brought Liberian men to this destination in gangs, groups of roughly twenty-five laborers from a common home region. They were contracted to the mines in a staggered hiring process described by the chief transport officer in the Gold Coast in the following manner: "They are enlisted in their own country informally, and the agreement is properly drawn up under the Master and Servant Ordinance on their arrival in this colony."[54] Acting on behalf of a specific mining firm, the Liberian labor agent, who spoke some English as well as the same language as the members of his gang, was the one driving the informal "collection" process.[55] As the transport officer explained: "A headman already in work [with a mining firm] gets from his employer a note stating he is authorized to assemble a gang of labourers. He proceeds with this to his own country."[56] The headman's employment relationship with the mining firm was cemented by enough trust for mine managers to advance him the capital and paperwork to pay advances to individual men, cover for the taxes requested by the Liberian government, and take his recruits on a costly voyage aboard a European steamer to Sekondi in the company's name. He received a small amount of cash from the mines in advance, and the balance was covered once they reached Sekondi. "On arrival at their destination the captain of the vessel calls upon the employer or his agent with a bill of charges for fares and for the head money paid by him to the Liberian Government prior to his sailing with the labourers."[57] It was not standard practice to dock workers' pay in order to cover these charges. However, it was not unheard of either. Some officials claimed that, "the cost of the steamer and the head money [for the Liberian government] is usually and properly paid by the employer; sometimes however, it is deducted from the men's wages."[58] As a next step, the work agreement was formalized. The men's term of service was almost invariably one year.[59] Then individual wage rates were determined. Though the recruiter may have given the men a rough estimate of their earning potential prior to leaving Liberia, nothing was certain before they endured a final physical inspection in front of mine managers or their agents:

"On being handed over to their employer they are informed of their rate of pay which varies with their apparent ability, and is usually reckoned at so much a month and not so much a day; an important distinction and one that is the pivot of numerous labour troubles."[60] Evidently, many laborers did not initially understand that they would be subjected to an extended pay system. In fact, they appear to have been rather vulnerable throughout the entire hiring process, from the obscure manner in which they were approached and made lavish promises by African recruiters working for the mines, to the necessity of being transported far away from their homes, to being contracted under the Master and Servant Act starting in the 1890s as largely illiterate individuals, and to having little knowledge about the precise conditions of their employment before starting work and later having difficulty leaving such a position. But it would be incorrect to paint the West African indentureship system as another form of forced labor due to the actions of these recruiters, especially taking into consideration how this relationship later unfolded. Because, generally speaking, these groups did not disintegrate when the young men were informed of their final wage rates. One barrier to the blatant and widespread deception by recruiters was the fact that mine managers by and large appointed these same agents to lead and train recruits on the mining concession over the course of the next few months. This motivated the agents to maintain their fragile bond with the recruits, at the very least to receive a positive reference from their employer. The labor agent, now turned supervisor (or "headman"), accompanied his men to Wassa and remained with them for the duration of their contract. Through all of these steps, from the point of contact in the home village or town, across rough waters to Sekondi, into the urbanizing mining centers of Wassa, and throughout their term of service, he was their immediate guardian and disciplinarian, their interpreter and on-the-job trainer.

But what inherent conflicts existed in these unequal ethnically governed, albeit temporary, bonds? Existing scholarship discusses the emergence and professional trajectory of African intermediaries in colonial mines in predominantly ideological terms. Following the theme of "colonial ethnicization," several scholars have reiterated that "tribal authority" fitted into contemporary European ideas about the backwardness of African society, or the "African worker" more specifically. Infusing a sense of "tribal" orientation into the organization and treatment served the purpose of relegating the obligation of supervision and caretaking to African leaders propped up by white managers who personally had little to no contact with Africans at the lowest level of the mining hierarchy. Thus, for many social groups coming to the mines for work, a vague and malleable notion of ethnic identity may have preceded the miners' term of service. However, the mines' bureaucratic structure further intensified ethnic classification, ingraining it not only into the professional but also into the social lives of its African laborers. Layers of proof support this perspective. Speaking of indirect recruitment for the Enugu colliery during the early colonial period, Carolyn Brown has put forward that both "the [colonial] state and the [colonial] industry used men whom they designated as local 'leaders' to control their populations."[61] This "local method," as she refers to indirect recruitment, "was adopted under the assumption that requests through 'native chiefs' would be accepted by their subjects who, it was alleged, were accustomed to serving their customary labor needs."[62] Illuminating one aspect of the colonial contradiction, Carola Lentz contends that labor migration to and work in the Gold Coast gold mines functioned as a civilizing project of sorts, but mine management together with colonial officials also aimed to retard rapid social change by imposing social control by means of ethnic belonging, though she goes on to explain how the workers themselves further reinforced these categories.[63] In looking at company records from this period, it quickly becomes evident that mine management supported the organization of migrant workers along ethnic rather than occupational lines. Employers even utilized ethnic identity as a predictor of a social group's productive value, physical strength, and general loyalty,[64] thereby turning ethnic self-identification into a variable in the attainment of wage work, which also had the potential of affecting wage rates. In addition, housing was organized along the lines of social background. The Lagos *village*, the Hausa *zongo*, and the Jato *zongo* are a few terms describing individual settlements for migrant labor in the southern Gold Coast. Some of these areas even had their own elected institutions, such as a "native council" and an "ethnic presidency," by the turn of the century.[65] The intention here is not to denigrate these arguments so much as to expand them by arguing (1) for the existence of a multiplicity of differently motivated labor agents in a given labor market; and (2) for more attention to be paid to the economic (and not just social-cultural) exchanges and negotiations that informed the gang relations. A more nuanced reading of the actions and reactions of mine managers, recruiters, and recruits helps to construct a more dynamic portrayal of the wage labor market in early colonial West Africa.

Economic Pragmatism and the Origin of the Corporate Indigenous Supervision System

The relationship between the companies and their indigenous labor agents is one part of the story that needs to be revisited. In his study of the Wassa mines, Jeff Crisp quotes J. R. Dickinson's assertion that for managers in the Wassa mines, "the easiest and most satisfactory, way of dealing with Africans" was through African authorities.[66] Although a lot of room remains for interpreting terms such as *satisfactory* and *easiest*, what shines through in this sentence is not necessarily just the managers' outlook on African backwardness, but also a need for pragmatism. Arguably, pragmatic interests were an important motivating factor in the formation and spread of the corporate indigenous supervision system, with some of the most basic of these concerns including: European employers' ability to deal with multiple language barriers in management, keeping the costs of white manpower to a minimum, limiting the companies' direct social obligations to individual laborers, and demonstrating compliance with the colonial system of indirect rule to ease relations with the administration.

Certain scholars have estimated that corporate indigenous supervision emerged in Liberia as early as the eighteenth century.[67] For the nineteenth century, there is rich documentation of small groups of young men from the eastern region of the state being hired through local authorities to work aboard ships as crew members. According to Captain William Allen of the Royal Navy, their contract was generally for a limited one- to two-year period during which the men were in charge of loading and discharging cargoes. In his travel diaries, published in 1841, he reported: "Both men-of-war and merchant ships take a gang on their arrival on the coast, for the purposes before mentioned [working the vessels]."[68] He recalled his own interactions with hopeful Kru boatmen at Cape Palmas in Liberia and on the Kru Coast in Sierra Leone during the 1830s. Allen had managed to procure one hundred men for the various departments of the expedition. Most were Kru men who would assist in steering the ship, in keeping with the popular belief that Europeans could be exposed to the sun only for limited periods. Yet a letter between the colonial agent of the American Colonization Society and the first governor of Liberia, the American Thomas Buchanan, following the establishment of the Commonwealth of Liberia in 1838, paints a more obscure picture of how the men were engaged. On October 22, 1838, the agent sent the governor a list of forty-three men (including a few "captains" and "lieutenants") who had been engaged to work aboard two vessels "with the distinct understanding that their attention may be called to various services, to fight or to fortify and perhaps both."[69] However, although the men had received cash advances and were to be paid up to the time when they were returned home by ship, their contracts were not fixed and they could be terminated at the employer's whim. As the agent explained, they "are engaged for no definite time—the term of their services remains entirely with you."[70] Finally, it is relevant that these gangs did not always come from labor-abundant pockets on the coast, waiting to be called onto a given ship in preorganized groups. Recruiting was a difficult task. Speaking of the Kpelle people, the agent complained: "They are thinly scattered over our whole (eastern) front, and could not be collected together in less than a period of days."[71] Therefore, local labor agents eliminated some of the time and revenue associated with finding, engaging, and managing a large number of wage earners in what was a land-abundant context.

A crucial first step for foreign actors was to engage a recruiter with a good rating from past employers and a long-term career outlook. Captain Allen emphasized the importance of obtaining references from employers: "It has not been considered necessary to give the Kroomen double pay, but to have the power of giving them better ratings than they usually have in men-of-war on the coast of Africa, to ensure getting the best description of men, as there will be no inducement of prize-money."[72] These were the limits of checks and balances in place before a container of sterling coins and lots of faith were placed in the recruiter's hand. On being formally hired, the labor agent became responsible for attracting, screening, and hiring reliable unskilled laborers. Under ideal circumstances, his preestablished socioeconomic networks would give him access to sources of voluntary labor that were out of the reach of most foreigners. These same networks were also critical when it came to screening potential candidates. In sum, foreign employers were spared the arduous and petty negotiations that weeding out the weakest recruits would have necessitated in what was to them a foreign environment.

Starting in the late nineteenth century, during a period of rapid economic expansion in West Africa, this form of recruitment spread equally fast into other parts of the region. Indigenous recruiters became indispensable intermediaries to foreign employers along the West African coast. As shown in the previous chapter, practically all of the African contract workers in Wassa during the last quarter of the nineteenth century also came from Liberia. Mine management expected the career-minded Liberian labor agent to remain loyal to his mining firm over several years while the larger population of laborers largely remained in flux. They took the lead in accompanying laborers to work sites from distant locations. It was without a doubt the faster but also safer option simply to cover the gangs' travel costs, because those locations might conceal hidden health hazards to the European outsider. Furthermore, recruiting through indigenous labor agents was certainly the most profitable path for employers in cases when these costs were eventually passed on to the individual laborers themselves.

Indigenous recruiters for the Wassa gold mines were usually men who had previously worked in the mines and had successfully adapted to life in this new environment. First, this experience gave them more credibility when trying to convince new men to migrate. Second, their presence simplified routine supervision. Though contracts under the Master and Servant Act of 1893 were introduced for Kru laborers during the 1890s, individual contracts did little to lighten the burden of keeping track of the labor force. Mine managers struggled to achieve a precise overview of the individuals in their employ at a given time in a given place, and the difficulties of labor supervision grew proportionally with the number of African contract men put to work. In 1905 in neighboring Asante, for example, a mine manager noticed to his own dismay that "there was a much larger number of natives employed [there] than [they] had any idea of owing to the system of contracts whereby only one name appeared on the books."[73] Frequent and widespread turnover on an annual or half-yearly basis only exacerbated the difficulties faced by Europeans who were trying to gain oversight over something as fundamental as the number of their African staff.

Entrepreneurs, officials, and private citizens in the colony frequently extolled the merits and superiority of white management in the mines. For example, in 1903 the chief officer of the government transport department contended that "the value of a gang of labourers varies with the ability of the European in charge. . . . One will condemn a gang as worthless, while another will get most excellent results out of it."[74] In this context, Europeans essentially had the answer to the labor question at their fingertips, namely by their own purportedly efficient supervision: "Probably under efficient supervision, or if possible working on a fair system of daily task work, a day's work will be performed that will for the expenditure compare favourably with that of an unskilled labour in other countries."[75] Similarly, visiting Wassa in 1905, Decima Moore Guggisberg believed that the potential of the

African laborer depended on "a firm but just hand."[76] In her eyes the African worker "is certainly cunning, imaginative of injury, and indisposed to physical exertion, but when properly treated and led by men who understand him, his working power compares more than favorably with that of other native races."[77] In 1909, the British miner W. J. Loring, who had worked in Wassa, related that "under good management an improvement can be effected [in the African labor force] and the standard raised."[78] Loring also declared, "It is not always easy to manage white labour, but to keep the blacks up to the mark even more experience and tact are required."[79]

In spite of their praise, however, all of these commentators had to face the reality that white labor was especially expensive in West Africa. In order to secure competent miners from Britain, the United States, or Australia, employers often had no other choice but to outbid the wages offered on other gold fields around the globe. The reasons behind these exorbitant costs were multilayered and also interconnected. One element adding to the general unpopularity of West African sites was the high probability of contracting a potentially deadly illness. With most men hoping to avoid the region that had earned the chilling title of the "White Man's Grave," a visitor to the Gold Coast reported that after signing a contract with a West African mining firm, white miners received "more money in their pockets than they had ever had before in their lives."[80] Attractive benefit packages accompanied these generous salaries. As one colonial observer testified, there was even gossip of desperate mining companies giving white miners "advances of pay, sometimes even guaranteed their bar bill" during the passage to West Africa—all in the hopes of getting them to overcome their apprehension.[81] One unintended consequence of this practice was that even men with mediocre mining skills suddenly found themselves in an advantageous bargaining position. "The country had an evil reputation for fever and death, so these men in the low state of the labour market which existed at the time, made most absurdly good terms for themselves with the companies at home."[82] As one colonial observer testified: "I've seen rough workmen, who had no more skilled knowledge of gold-mining than what they had acquired as colliers or tin miners at home, coming out to direct work requiring great technical knowledge, drawing £60 to £40 a month with food found, and . . . first-class passages provided."[83] Therefore, it was hardly surprising that managerial problems persisted even after miners from Britain, Australia, or America had agreed to work there. General managers complained about the low productivity levels of these men compared with what was expected on gold fields elsewhere. It is not clear whether this was mostly due to the employment of less qualified men, or if environmental and health issues were to blame. Both of these issues certainly factored into the lamentation of one mining entrepreneur in Wassa that entering into a contract "with Europeans, is most unsatisfactory to the manager."[84] It was reported that from "10 to 15 percent more officials are needed to do the same work here than in other places."[85] Furthermore, although the overall quality of white miners may have improved significantly between the nineteenth and twentieth centuries, colonial commentators still assessed the majority of them as being largely "incompetent" in terms of output when compared with other mining centers.[86] It perhaps does not need saying that the health hazards of West Africa did affect white labor in a real way, with real negative impacts on production rates in the mines. The frequent bouts of incapacity that white miners experienced was "not on account of holidays, but on account of sickness."[87] The forest environment around Tarkwa, a district situated "in the midst of a dense, damp, and malarious [sic] forest," was objectively harsh.[88] Not being entirely convinced that it was not the environment alone, the Australian surveyor W. L. Crompton put forward that the "continual high moist temperature is in itself a severe strain on ones nervous system." Then there was the lack of available medical facilities. In his view, "being so utterly cut off from outside help makes it worse, and the extreme suddenness of disease here makes it unwise to contemplate illness, yet, at the same time any carelessness lets you know at once that malaria is always lying low for you."[89] According to another mining agent: "The indirect losses, due to decreased vitality, in preventing the best engineering talent from seeking a career in West Africa, and in frequent changes of staff, are difficult to assess in terms of pennyweights, but they are considerable."[90] Financial losses on white labor due to illness and recovery improved slightly between the nineteenth and twentieth centuries;[91] however, management continued to struggle. In 1901, the director of a local mine at Fanti notified shareholders of an endemic state of affairs: "A certain proportion of the men are laid up for a few days or a week at a time with a mild form of malarial fever, and that is no doubt likely to continue, though we hope now that we have sufficient capital and can afford to do more clearing work and to build better houses and provide better living for the men, that there will be an improvement in the health of the mining camps."[92] Raising the living standards of whites at the Fanti mines closer to the level of those living in Accra was necessary. Visitors to both of these settlements could not help but point out the remarkable gap in living conditions:

"Government officials have twelve months' work, six months' holidays on full pay, they live largely on the coast and have good bungalows. Miners must live in the dense bush and swamps in the interior without good bungalows, often tents only."[93]

Although the previous quote implies that the holidays of white miners were shorter than those of colonial officials, their length was still a cause for concern for mine managers. Somewhat ironically, the prevention of illness encouraged further disruptions to management. To keep their men in a relatively good state of health, prominent properties such as Abbontiakoon and Adjah Bippo relied on rotating field management.[94] White workers, whether in administrative jobs in accounting or technical jobs such as surveying, were generally engaged on an annual basis. Out of the twelve months however, they actually resided in West Africa for only around eight to nine months.[95] The last third of the year was spent back home, recovering from the ill effects of the climate.[96] Long holidays, illness, anxiety, fatigue, and West Africa's reputation obliged mining entrepreneurs to face the dire fact that if a mining company hired eight workers, it would lose "on a rule thirty-eight days [of work] a month."[97] In 1901, the directors of the Fanti Consolidated Mines confirmed, "The worst that can happen to us, looking at it from £ s. d. point of view, is that we shall have to spend 25 per cent more on our white staff. . . . Well, 25 per cent added to the cost of white labour means 6 ½ added to the total cost of raising and milling the ore. That is the worst that we have to expect from the climate."[98] Nevertheless, informed directors and managers, forced to grapple constantly with this dilemma, mostly downplayed it in front of their shareholders.

Census reports consolidated in table 2.3 for the number of European and African laborers employed in the mines between 1900 and 1918 paint a corresponding picture. In 1904, 611 European worked alongside 17,044 African miners, at a ratio of 1 to 28. Both populations declined in 1905, leading to a count of 504 Europeans to 12,465 Africans working in the mines. In 1906, the ratio improved slightly, with 540 European to 14,760 African mine workers. Nevertheless, especially taking the habitual mobility and disability of large sections of white miners into consideration, it is safe to say that they constituted not much more than a "thin white line" that relied on African intermediaries in order to function. In the nineteenth century, for the roughly 200 to 400 Africans laboring in the extraction and hauling of ore, or across the various mills spread across a given concession, or in the forests collecting firewood, there were about two white mine captains under the direction of a chief mine captain with the duty of supervising their

Table 2.3. Labor employed in the mines, 1900–1918

	Average number of African laborers employed daily by the mining and dredging companies	Average number of European laborers employed daily by the mining and dredging companies	Overall number of European residents employed by the mining and dredging companies
1900	—*	—	—
1901	—	—	—‡
1902	—	—	778
1903	—	—	1043
1904	17,044	611	1222
1905	12,465	504	1157
1906	14,760	540	992
1907	15,277	—	883
1908	15,796	—	759
1909	15,895	—	585
1910	19,138†	—	660
1911	19,153†	—	922
1912	17,633†	—	953
1913	15,658†	—	928
1914	15,741†	—	1020
1915	15,300†	—	481
1916	15,296	—	637
1917	16,004	415	714
1918	13,918	336	578

* No data available
† Number of European residents not recorded before 1902
‡ Includes Europeans
Source: Annual Reports for the Gold Mining Industry 1904 to 1908 PRO CO98 14–16; Gold Coast Departmental Reports for the Years 1900 to 1918, Digitized Books from the University of Illinois at Urbana-Champaign and the Open Content Alliance, http://hdl.handle.net/10111/UIUCAFRICANA:Serial/5530214.

activities. The surveillance and training of individual gangs of African laborers by Europeans was made even more perplexing by cultural barriers. Since few European officials or individuals in the private sector had the time or will to learn even one of the many languages that would have allowed them direct communication with extensive African populations, middlemen had another important role to play. Although European employers in colonial offices and courts often relied on educated African middlemen who sometimes spoke multiple European languages as well as indigenous ones, the diversity of the urbanizing mines made this issue even more complex. Employers therefore required that African supervisors "must know [a] little English"[99] to help them interpret among the various social groups on the property. Mine managers also profited from the skills that these indigenous authorities passed on to the inexperienced recruits. After the recruitment procedure, once his gang was formally taken on by a company, the gang leader's responsibilities shifted to mining production. However, he was freed from the drudgery of manual labor so that he could focus on educating the members of his gang in how to conduct mining work in an efficient manner. The range of his duties also extended beyond the workplace to include settling minor legal disputes among workers.

Indeed, the indigenous supervisor took over many areas of social protection that the companies themselves were reluctant to regulate. Negotiating minor legal disputes among members of his gang naturally put laborers in somewhat of a bind when they experienced abuse by their leaders. Beyond mediating among them, he was also allowed to exact small fines, which seldom exceeded five shillings; collecting fines would not have been all too difficult since wages, too, were dispersed by the gang leader. Legal infractions could include anything from unfinished work to nuisance disturbances. In cases of desertion as well, the indigenous supervisor was the first in line to search for and to collect runaways. Only when deserters were found but refused to return to work were they brought before a district court.

Power Relations between Indigenous Supervisors and Recruits

If the system of corporate indigenous recruitment in Liberia initially relied on the desperation of a few young Kru men to build a better life, as early as the 1840s it had become a mainstream phenomenon. William Allen attested that it "has now become so general among them—as the only way of acquiring riches—that nearly all the male population spend a shorter or longer term of probation, either in trade ships or in vessels of war."[100] In addition to being widespread, gang labor, which he described as a sort of apprenticeship, had become aspirational. This was illuminated, for instance, in the high status and honor that the icons and rituals of work life in that specific environment imbued on their wearers and performers—so much so that the uniforms and rank devices of deceased officers, acquired at local auctions, turned into objects of desire. The following observation was made in Kru Town in Sierra Leone, a temporary residence for laborers until "their service may be required":[101] "Being always the highest bidders at the sale of deceased officers' effects, articles of uniform are purchased by them; and it is not unusual to see one with a post-captain's coat and epaulettes, surmounting a waistcloth; or another with a scanty fold of cotton round the middle, while his head and the lower extremities are severally encased in a cocked hat and pair of Wellington boots."[102] These items were worn on special occasions and in reenactments. Allen recalled, "The lucky possessors of such outward insignia of office, are fond of imitating the routine of a man-of-war, in Kru Town, by mustering at divisions, the officer of the watch, spy-glass in hand, reporting to the Captain, etc."[103] However, according to Allen, this performance moved beyond entertainment to function as a form of appropriation that ultimately made attractive this path of employment: "This produces emulation and a desire to serve on board ship, and to merit a 'good book' or character."[104] This consideration ends up being particularly valuable for the manner in which it distinguishes the recruitment practices for wage work from what happened to slaves, slavery not being illegal in Liberia until the 1930s. If recruits had come from a population of slaves, what is the likelihood that recruiters would have gone to such lengths to make these routines of work aboard a trade or naval vessel seem more attractive? Instead, the anecdote highlights the wooing of men with a large degree of free choice, the promise of honor, and a better future looming large in such performance. Granted, gang labor was not entirely free of all deception. Recall that the finer details of work conditions and wage rates remained unknown until an individual contract was signed at the place of employment. Yet by and large the recruits seem to have been making an active choice in favor of this work in spite of potential pitfalls.

The agency and power of laborers in the gang system has generally been discussed through the lens of their relationship with the leaders of their gangs, the headmen. Photographic evidence and other anecdotal sources confirm the existence of great social disparities between the labor agent and his recruits. Differences in social stature were immediately visible on encountering them. Cloth, as Jean Allman has shown, "became a ready index of the extent of capitalist penetration" during the early years of colonial rule, distinguishing the clothed, and therefore "civilized," individual from the naked, and therefore "backward," one.[105] The gang leader was usually seen dressed in Western clothing, the cotton T-shirt, full-length trousers, and cloth cap so typical of a contemporary industrial worker. In contrast, typical unskilled workers frequently wore loincloths and traditional-cloth garb, as reflected in the aforementioned performances. Such blatant material inequality makes it that much easier to imagine that this system was prone to abuse, as does the obscurity surrounding the negotiations that brought them together in the first place. Therefore, the following sections are deliberate in their attempt to distinguish a deeply abusive and exploitative situation from one that was essentially precarious and facilitated abuse. They underline the point that just because the indirect recruitment system remained largely outside the purview and control of Europeans, that did not necessarily make it dubious. To some degree migrant laborers benefited from and therefore upheld a system by which these agents provided them with knowledge, services, and social protections that were essential to their stay and survival in Wassa. Driven primarily by the need to earn wages and cash advances, migrant laborers also found themselves in a strong bargaining position at the beginning of their term of service, especially in the context of the irregular implementation of labor laws. Nevertheless, the negligible involvement of the colonial state had the effect of making migration through these informal mobility agents a risky endeavor overall. Certainly, after the first few months of service, recruits were left in an increasingly vulnerable position. The very last section of this chapter will analyze laborers' motivations through their own correspondence, emphasizing how they discussed the costs and opportunities associated with both migrant (gang) labor and work in colonial commerce in general. Although scholars studying indentured laborers in other historical contexts have found the opposite to be the case, we will see that economic opportunity, much more than economic coercion or physical force, seems to have motivated contract migrant laborers on the southern Gold Coast.

Figure 2.2. Kru labor agent (*top left*) with other workers in the southern Gold Coast Colony, ca. 1876–77. The National Archives UK, CO 1069-29-35. Wikimedia Commons.

Workers' Politics and Welfare

In Carola Lentz's research on ethnicity on the Gold Coast, she insists that although structural and administrative pressures imposed a new layer of rigidity onto ethnic identity, this change was further reinforced by those subjected to such categories, the migrant laborers themselves. For these young men, life in urban areas was both invigorating and terrifying, and ethnicity often held the key to new means of empowerment. Although for many of them ethnic identity had not been of great, or perhaps any, significance during the precolonial period, they now viewed it as an asset because it produced "new forms of classification and self-understanding which differed considerably from precolonial models of belonging."[106] The indigenous recruiter-supervisor was part of this novel framework, which was meant to meet the needs of a "tribal" people in a cosmopolitan environment but also had an equalizing effect on intra-African relations. In spite of any historical tensions, competition, and imbalances of power, in the urbanizing mining centers groups from different localities encountered one another on a more or less equal footing.[107] In a mirroring of the transformations that were happening on a broader political scale, ethnicity gave laborers new tools for self-improvement—new liberties, rights and powers relative to other social groups migrating to the area.[108] These "new political structures, rituals and discourses were attractive . . . to migrant workers . . . who sought the same status and respect as colleagues . . . from precolonial kingdoms."[109] To name one of the advantages of this new statehood, all casual and contract workers, regardless of their social origin and background, now received (almost) equal pay for equal work. Moreover, they were affected and protected by labor laws in practically the same manner, notwithstanding the judicial authority given to their individual gang leaders. For this reason, among others, migrant men accepted and affirmed affiliations that could protect their interests along lines that were not occupational but that had ethnic qualities, though to be clear such affiliations did actually extend beyond the confines of individual rural villages. They consciously imposed these classifications onto themselves by congregating with others with whom they shared commonalities, such as language, cultural habits, and often religion. The system of indigenous supervision in the mines, in particular, was known to provide benefits in important areas of social security. In general, the gang leaders' supervisory and punitive responsibilities were offset by the other services they provided to assist the laborers during their term of service. Mine managers perceived and molded them not only as disciplinarians and supervisors but also as educators and welfare agents. The agent had to serve as the ethnic broker to the members of his gang in the new work and living space. Consequently, in addition to voicing the opinions and grievances of the laborers in moments of discontent in the workplace, he also had to iron out tensions with other social groups in and around the mining centers when work was out.

It was crucial to be part of a solid social network on entering into a contract in Wassa. Men occasionally hired themselves to the mines without such a representative and found themselves in a position that was difficult to navigate, especially once the system had widely taken root. One administrative official explained how occupational contact networks and friendships were essential to survival in the mining villages: "If a man did not get food he could not work, and if he were a foreigner and without friends to help him on the particular mine where he was, he must go and steal to keep himself alive."[110] His statements are especially relevant to the discussion on colonial ethnicization for the way that he distinguishes the intention behind administrators' desire to maintain strong ethnic units in the mining centers from the actions of supposedly oblivious mine managers: "I have had to put this very plainly to several mine managers, who, not knowing the country, did not recognize that their labourers were drawn from numerous tribes who would not sympathize with each other if in trouble."[111] In addition to making sure everyone was fed, the gang leader arranged accommodation for his men.[112] He was obligated to look after them in times of illness as well.[113] For their own convenience, managers in Wassa were explicit about the fact that sickness was not a reasonable cause for letting go of or replacing a worker. Ignoring this rule was an offense that merited severe punishment, as shown in a letter between a gang leader and an employer who accuses the gang leader of such misdeeds. The employer had recently learned of the leader's dismissal of a sick member of his gang, and this was not the first time something of this nature had occurred under his guardianship, either. In his defense and with the assistance of a local letter writer, the gang leader responded in slightly broken English: "In reply I beg to state, that I was very much surprised to hear that a sick boy went to you and complained to you that I had driven him from my premises and that I am not care of my sick, except when they are in position, and that I had driven two sick boys from my premises before, and after that if I had done so, you will soon dispense with my services."[114] To justify his actions, and most important, to keep his job and maintain his professional rating, the leader painted a picture of the laborer Supply, who had made the complaint, as a lazy and deceitful recruit. The letter is illuminating for its ability to demonstrate that not providing welfare services, such as "taking care of your sick," could quickly put a gang leader's reputation in jeopardy. Such a reputation was hard earned and embodied his sole means of gaining future work in this field.

Although the above observation in some ways confirms the pliable image of "career-minded" labor agents, eager to meet the expectations and carry out the orders of mine managers, perhaps there is more to consider. It has been acknowledged that these figures emerged in order to meet the needs of capital and that they fit into the ideological framework of foreign employers. However, there was much more to the motivations and strategies of these middlemen. The indigenous recruitment and supervision system was an important bridge between "traditional" and "modern" employment relations especially since wage employment remained largely temporary in nature. Therefore, like other intermediaries in colonial Africa, the labor agent was a complex figure who alternated between serving a European boss and "fulfill[ing] the goals and concerns of his people,"[115] though differences were certainly more pronounced in labor agents who were either more or less dependent on the mining firms to begin with. The scholarly focus on their "pliancy" and capitalist control tends to undermine the agency of African laborers, not to mention ignore the fact that the politics of labor exploitation largely continued to play out between Africans, beyond the purview of European managers. Although some of the existing scholarship on the Wassa mines has automatically taken on a perspective predetermined by classical Marxist theory, the negotiations that were happening between Africans on the mines provide material for a whole new set of questions and observations. For instance, details about the mechanisms of recruitment through indigenous labor agents show that in addition to displaying power in culturally appropriate ways in order to convince potential recruits to leave their villages to stay on a particular concession for an extended period of time, they needed to give the recruits strong economic incentives as well. In other words, in order to keep in balance this fragile and unequal power structure, the recruiters had to guarantee certain social and economic protections to the recruits, not to mention economic opportunity in and of itself. Surprisingly few historians of Africa have paid close attention to the monetary aspects of the gang relationship.

82 Chapter Two

Figure 2.3. Illustration of a gang of Kru workers in the 1890s. Alphonse-Jules Wauters, *Stanleys Expedition till Emin Paschas undsättning*, trans. Oscar Heinrich Dumrath (1890). Wikimedia Commons.

Wages, Cash Advances and the Spread of Calculated Debt Bondage

Some contemporary commentators attributed significant authority to the labor agents. For instance, a missionary stationed on the Gold Coast in 1909 insisted that Kru labor agents had "almost absolute power" over the laborers in their gangs.[116] After an encounter with a gang of Yoruba woodcutters residing in the Wassa area, another missionary referred to the gang leaders as "prophet-like figures," emphasizing their spiritual authority of these leaders.[117] These colonial observers unsurprisingly put great emphasis on the social power held by what they perceived as quintessential "tribal" leaders—a bias, which may explain why differently motivated labor agents were indiscriminately referred to as "headmen" by colonial observers. Yet, after a quick look at court records from this era, it becomes quite easy to discern the importance of economic incentives for recruitment. There are two key observations to draw from such materials. Firstly, with the larger economy of the Gold Coast and its protectorates taken into account, the wage rates offered to African miners around the late nineteenth and early twentieth century were actually quite considerable (compared with gains from work in agriculture, for example). Secondly, not only did it become increasingly commonplace for recruits to receive a cash advance, but also the amount they could expect to pocket grew over time. And this was not because of the calculations of litigious recruiters and employers, but rather mainly because potential laborers for wage work increasingly found themselves in better bargaining position during this period.

On the subject of wages, recent studies by scholars such as Ewout Frankema and Marlous van Waijenburg have found that real wages in British West Africa were well above subsistence level.[118] Indeed, they were significantly higher than incomes in major Asian cities, and they increased significantly over time. In Wassa in the late nineteenth and early twentieth centuries, this trajectory was no different. Skilled underground men often made twice the average daily income of rural agricultural workers for the period under investigation. The records of the Abbontiakoon Mines, shown in table 2.4, indicate that on average drill boys and shaftmen earned two shillings and an additional three pence for subsistence in 1903, and this amount increased automatically if they stayed on after six months of service. On average, surface miners at the same property earned one shilling three pence with an additional three pence. Other skilled laborers, including carpenters, blacksmiths, masons, fitters, engine drivers, and firemen, could expect to earn between two shillings three pence (with three pence for rice and other foodstuffs) and three shillings three pence (with three pence for food). These were competitive salaries compared with those the state set as a standard for the mining companies a year later, as seen in table 2.5. Thus, when in 1906 an Australian miner complained that "black labour is not cheap here," explaining that "the majority of boys working in mines average 1/9 [one shilling nine pence] per day,"[119] he was not just exaggerating in order to make an argument for giving even higher wages to white miners. As the traveler Mary Gaunt noted, this was because the African worker "does not like working underground, for which few people I think will blame him, therefore high wages have to be paid."[120] It also has to be remembered that in a racialized colonial context, an enormous gulf in pay persisted between the real wages of white and black mine workers.

Only a few companies actually managed to orient themselves to what the colonial state had decided were fair wage rates, but actual undercut the market rate. And actually, when employers did fiddle around with the pay books, the goal was not usually to further depress wages. The trend was in the opposite direction. Under conditions of labor scarcity, they attempted to devise new ways to attract laborers; in doing so,

Table 2.4. Average daily wages paid at Abbontiakoon mines, ca. 1903

Carpenters	3s 6d	
Blacksmiths	2s 6d	
Masons	2s 6d	
Fitters	2s 6d	
Engine drivers	2s 6d	+3d rice
Firemen	1s 9d + 2s	+3d rice
Painters	2s	+3d rice
Drill boys	2s	+3d rice
Shaftmen	2s	
Shaftmen after six months of service	2s 6	+3d rice
Timbermen	2s + 2s 6d	+3d rice
Miners	1s 6d	+3d rice
Fanti laborers	1s 3d	+3d rice
Krepe (Ewe)	1s 3d	+3d rice
Tramming in mines	1s 3d	+3d rice
Kru Laborers	1s 3d	+3d rice
Senegalese	1s 3d	+3d rice
Not paid on Sundays; not paid when sick.		

Source: Letter from James Bullen the Hon. Secretary of the Mine Managers' Association, Abbontiakoon, May 11, 1902, CUL RCMS 139/12/35.

Table 2.5. Daily wages paid to laborers in the mines in the Gold Coast and Asante, 1904

Miners	from 1s 6d
Shaftmen	from 1s 9d
Surface laborers	1s 3d
Fitters	3s to 3s 6d
Carpenters	from 2s 6d to 3s 6d

Source: Report on the Mining Industry for the Year 1904, PRO CO 98/14, p. 5.

they often clashed with competing concessions. In order to avoid having to raise the total sum of wages, some employers experimented with manipulating the ratio of extended wages (usually paid after a number of months) to daily subsistence payments. For example, in 1903 the manager of the Abosso mine began to offer men who were already earning one shilling in wages and three pence, nine pence daily pay and six pence subsistence instead, as a means of retaining more laborers. The idea behind the plan was that a greater share of cash in pocket (until the contract was over) would help to motivate laborers. Having additional cash could certainly ease a miner's stay at the mines by, for example, lessening the likelihood of his falling into debt with a moneylender who would force him to prolong his service in order to pay of any incurred interest. Thus, although it was important to make a good daily wage, the extended payment system gave even greater value to any money that could be immediately accessed and put to use. The end result was that among African contract men, as one observer noticed in 1903, "£2 a month with an advance of 20s [shillings] is more desirable than £2 10s [shillings] a month with no advance."[121] Furthermore, in a break from the past, by 1903 it was not just the migrant contract men who received advances but even contract men recruited in the coastal regions of the Gold Coast and its protectorates.[122] In its original form, the cash advance was meant to cover a migrant laborer's most basic needs before he left his home. In 1838, a contractor in Liberia explained his reasons for giving men advances as (1) to "subsist their families"; and (2) so that they could "obtain some little necessaries for the campaign."[123] As a result, the actual amount of the advance could vary significantly from one laborer to the next. On the Gold Coast, there was initially logical reasoning behind the advance payment going out to African contract men as well, though it was not the need to support one's family that made the strongest argument in this instance. Advances to mineworkers were supposed to free potential recruits from debt obligations, that is, creditors, back in their rural homes. According to one administrator, foreign employers were willing to hand out cash advances "primarily for the purpose of paying his [the laborer's] debts (not for supporting his family), as otherwise his creditors, and natives have creditors, would not allow leave."[124] Hill has made a convincing case that credit relationships were already an integral part of rural based economic production and development in parts of West Africa going back to "time immemorial."[125] She put forward that giving credit was, not simply a means of profit making. There was the unspoken rule that well-off farmers would provide loans to poorer ones, tying them together in a loose web of mutual support. The legal end of slavery in 1875 left a void in many social and economic networks. As a result, new types of dependencies, rooted in debt-based relationships, were formed and multiplied, eventually becoming the norm in many areas. Furthermore, as this study demonstrates, credit relationships or dependencies almost immediately began to shape and facilitate economic relations in urban areas as well.

What may have started as a sort of relief program for new recruits for contract work quickly morphed into something more unruly. By and by, individual workers could expect to receive an advance payment even when they were not really experiencing any particular form of financial hardship. Advance payments were gradually integrated into the recruitment process solely for the purpose of enticing workers. In consequence, men who had agreed to work in Wassa on a contract basis were regularly promised between ten and thirty shillings upfront, the equivalent of between roughly eight days' and one month's pay for an unskilled laborer. It primarily became a means to get men to agree to contract work. Expressing his concern over this development, one transport officer stated, "If men are engaged at their homes the advance system is reasonable, but men at the mines get advances far from home and for no reason other than as an inducement to take work."[126] Many colonial authorities operated by a moral principle that determined that the spread of credit needed to be curbed, ignoring the economic and social logic of this provision from the perspective of the competing mines or the labor agents. But handing out these advances unselectively did put them in a vulnerable position, especially given the limited physical or legal pressure the mining companies could exert on contract men in order to ensure their service at the time.

Loose Labor Regulations and the Advance Pay System

In a 1906 report for the British Parliament on "Coolie [*sic*] Labour" in the British Empire," Herbert Bryan, the acting governor of the Gold Coast, stated that "labour [in the Gold Coast] is seldom 'indentured' in the strict legal interpretation of the word."[127] Although contracts and cash advances were as much part of the recruitment process in the Gold Coast as in other parts of the British Empire, the system put into place there was much less rigorous. The governor's words outlined a framework for labor recruitment with very loose legal and physical boundaries. It presented a particularly troubling state for mine managers and recruiters who had fallen into a pattern of offering high cash advances to new, potentially unreliable recruits.

As often was the case for Asian and Indian indentured laborers, for those in Wassa "bonds of community were backed up by bonds of debt."[128] However, labor laws in the Gold Coast of the late nineteenth and early twentieth century were neither written nor implemented to maximize the disciplinary powers of employers and recruiters, enabling them to easily instrumentalize such debt to their own benefit. A master and servant ordinance was passed in 1877, being drafted in 1875 to complement the Emancipation Proclamation of 1874. Parliament imagined it as a tool for eliminating the indigenous slavery system on a case-by-case basis, because a slave could use it to challenge the nature of his or her master's rights and power. Individual contracts for wage work were also thought of as the "modern" alternative to this archaic system. Isaacus Adzoxornu has observed that it was also meant to "fulfill an educational role" for a people who, it was presumed, were not well accustomed to hired labor, if at all.[129] Adzoxornu contended that a third intent behind the introduction of individual contracts under the Master and Servant Act was colonial paternalism, perhaps connected to the previous point. Even though individual contracts under the Master and Servant Act were not explicitly meant to benefit foreign employers, there was an understanding that the law could and, under ideal circumstances, would progressively serve such a purpose. Staff members of the Colonial Office discussed how, if labor "in the Gold Coast should become, as we hope it may, much more productive and profitable than it is at present, the inevitable consequence will follow that it will be extorted in large measure and by severer means."[130] And indeed, over time, amendments were made in favor of further disciplining African workers, for example, by allowing longer contractual terms and stiffer criminal penalties. However, such changes were much slower to inform laws in the Gold Coast than in those African colonies where the state closely identified with industry in the nineteenth century. Early versions of the Master and Servant Ordinance for the Gold Coast appear to have been rather mild. For instance, a cursory glimpse at the Master and Servant Act of 1893 demonstrates that employers and employees were equally accountable to the law for breach of contract.[131] Indeed, the ordinance actually stated, "Whenever the employers or employed shall neglect or refuse to fulfill a contract of service, or whenever any question, difference or dispute shall arise as to the rights or liabilities of either party, or touching any misconduct or ill-treatment of either party, or any injury to the person or property of with party, und any contract of service, the party feeling aggrieved may make a complaint to the court, which may thereupon issue a summon to the party complained against."[132] In short, in the Gold Coast of 1893, both parties were still equally liable for a breach of contract under civil law. Over time, however, administrators did slowly comply with some of the employers' protestations by using criminal penal measures to control labor, ending in 1931.

The colonial state's reluctance to impose new, stricter laws as a means of controlling African workers in Wassa is illustrated in chapter 4, through the story of the government transport department. Although the state agreed to help bring more contract workers to the mining industry, the chief transport officer's petition for the legal system to develop mechanisms to force more workers into continuing their service, or be severely punished for leaving, was highly controversial. The Accra administrators' overall desire to engage with labor matters on the concession was minimal. They therefore continued to sell the idea of the individual contract as *the* fundamental institution regulating relations between employers and employees without assuming a greater role in upholding the rules. Indeed, official records for the district surrounding the mining centers demonstrate quite clearly that cases of breach of contract only rarely ended up in colonial courts, in spite of endemic rates of desertion during the period under investigation. Although certain scholars have praised individual contracts under the Master and Servant Act as contributing to "the development of the norms and praxis of industrial relations" on the Gold Coast, additional attention ought to be directed to the historicization of law enforcement by the colonial authorities when assessing the value of legal instruments.[133]

Deserting before paying back advances already had become prevalent by the 1890s, so much so that by 1902, following pressure from the Mine Managers' Association, it became an offence punishable by six months' imprisonment with hard labor. Yet in spite of the theoretical power of contracts under the master and servant act, when put into practice the law often failed to hinder or deter laborers from absconding. For the most part, the option of leaving without notice was only really taken off the table once a laborer had paid back his advances and the company was in his debt—that is, a month or two into his service, making the preceeding period especially uneasy for the company and its recruiters. However, although it is evident that absconding occurred with greater frequency during the first month or two of service, laborers were not necessarily out to cheat the advance payment system.

Court records from Tarkwa District illuminate a variety of conflicts and contractual disagreements that could lead to desertion.

It is not clear whether the courts provided interpreters for the defendants, or whether these workers used their own versions of West African pidgin English, picked up during their daily work interactions on the coast, to make their cases. Either way, the grasp of the language they demonstrated during such trials leaves one to question the extent to which language itself could be the source of conflict at the colonial workplace. Is it possible that different parties differently understood the most basic orders and principles of their agreements? These court cases, related as they were to labor disputes in the mining centers, are useful for their ability to shine a spotlight on the mores subtle realms of labor negotiations at colonial mines. Here, they will be used to discuss the diverse causes and effects of desertion. To start, it was not unheard of for a recruit in the rural area to accept an advance and even have his transportation to Wassa paid for, only to disappear into the urban landscape in order to pursue other employment opportunities. This happened in 1909 in the case of the contract laborer Englishman, who after receiving a one pound advance and fifteen shillings for subsistence from the Abbontiakoon Mines, suddenly fell ill of jaundice, according to his own testimony. After recovering, Englishman then went missing from the concession for a two-month period, at which point the company sent his gang leader to find him. Englishman was eventually discovered "working in the railway yard clearing grass with a cutlass." He later admitted to having been discontented with work conditions on the property, saying it was too difficult. When Englishman refused to return to Abbontiakoon, he was brought in front of a district commissioner's court.[134] A similar case is documented in the legal trial of Kwesi K., who was accused of defrauding an entire eight pounds twelve pence and a railway ticket from a mining company via one of their labor agents.[135]

There were other financial matters to consider in addition to the cash advance problem. In particular, the additional channels of credit that recruits could access during their term of service were far more troubling issues to the gang leaders at least. Laborers depended on these sums because of the nature of the extended contract system. Although they could occasionally secure small loans directly through their gang leader, to his dismay, they also could accept a loan from a local moneylender. Third-party creditors provided these loans, with the understanding that ultimately the workers' supervisor-recruiter would be responsible for making sure their money was returned. Thus, the gang leader could quickly be put in a bind without even having given his initial consent. A 1910 letter by the indigenous supervisor Bowee reveals how anxiety-ridden men in his position could become as a result. Bowee led one of the many gangs that the transport department contracted out to mining companies for a fee, starting in 1903. Writing to the chief officer of the government transport department, he complained about the financial burden that his recruits had placed on his shoulders: "I have been taken all debts, whenever they take any credit from the Fanti people[136] and go up when they signed without notice [without the gang leader's approval]."[137] Conveniently ignoring what he stood to gain from this arrangement in the long run, he added that he was "not receiving anything from them" and was "doing this all on your account for [*sic*] the respect I am always getting from you [the chief transport officer]."[138] In the end, Bowee's objective was to get a good rating by maintaining his reputation with his employer, thereby improving his prospects for future work in the wage-labor market. The possible gain warranted the risk of financial strife.

The significant number of recorded cases of laborers who bucked the system does not erase the fact that in other instances, employers used the advance system as a coercive tool. Accepting an advance stood as a ratifying seal of approval on a laborer's consent to serve. Therefore, among the few trials for breach of contract under the Master and Servant Act located in the Tarkwa district records, debt was occasionally named as the primary factor hindering a withdrawal of labor, allowing the court to enforce the fulfillment of a contract when laborers were found guilty. Alternatively, workers were sentenced to hard labor for the state. There was no other legal means for an employer (or the state) to compel a contract man to work. The instrumentalization of debt can be examined in a case put forward by Gerhard Stockfeld of the Abosso Gold Mining Company in 1904. He listed debt as part of his legal argument against an absconded worker called Joe Mendi, who had signed on to work for twelve months on June 24: "On 22nd of June [he and other men] got advances. Accused left without motive on the end of [September] or early in October."[139] Following a similar pattern, in 1914 the superintendent of African laborers at the Abbontiakoon mines, Charles Robert Miller, reported a gang of deserters who were given advances and then sent down to the coast by rail by the company. Miller gave them work on the Cinnamon Bippo concession, "where they were given further advances [and] rice."[140] Unsure of their precise date of departure, Miller estimated that the men had deserted from the property between July 13 and August 1. They were all still in the company's debt at the time.[141]

It also has to be considered that it was exceedingly difficult for an unskilled laborer to get out of a contract by legal means once debt was a factor in the hiring process. For instance, on September 21, 1904, it was recorded that the contract men Illasi, Salakor, Abbey, Adessoman, and Hadji had left the Abbontiakoon mines without any intention of returning. Before doing so, when confronted by the mine manager John White, they offered to return their individual one pound advance payment to the company. After some of them had paid back the money, however, White compounded the accusation by suggesting that the men owed the company additional money for subsistence starting from the day on which they had been formally engaged. He had hired the gang at the coastal town of Sekondi just weeks earlier, on September 6. During the court proceeding White recalled meeting the gang of thirty-five laborers whose travel he had arranged from Accra, in Sekondi: "At Sekondi they entered into contract with me before the District Commissioner of Sekondi (contract produced)." They were later sent to Tarkwa. After producing the original contract, White described how the men had arrived at Tarkwa on September 12 and started work two days later, likely after having secured or built their own housing. However, on the twentieth "the headman [leader] of gang reported that one gang of 6 boys [sic] refused to work," without giving any definite reason.[142] The gang leader Illasi, who had worked for the company in the past, had accompanied the group all the way from Lagos to Accra. He attested to having previously told the manager that work conditions on the concession had caused most of the men to fall ill. This was the main reason for their leaving to find work in Abosso. His testimony was as follows:

> I live Abontiakoon. I know accused. They are my boys. I work with White. I bring them from Accra for White. He met us at Sekondi. I went. They signed on before Commissioner. We bring the boys to Tarkwa. They began to work. Yesterday [in the recent past] they refuse to work because it is hard. I reported it to White. He went to ask them to come back. They refused. They offered to pay money. They pay £1 7s 9d.[143]

The district commissioner required the gang to reimburse all expenditures on transportation that its members had accumulated during the recruitment process, for the journey between Accra and Sekondi, and for their travels between Sekondi and Tarkwa. Since "incurring debt under false pretense" was a crime carrying severe fines and months of hard labor,[144] it was not out of the ordinary for two of the five laborers, despite returning the advance payment in full, to be fined three pounds or one month of imprisonment. A third laborer, who had returned seven shillings and nine pence, was liable for the same fine. The remaining two laborers had to pay five pounds or suffer two months' imprisonment.[145]

The case of the African laborer Bokai, who succeeded in reimbursing the mining company that engaged him before leaving the concession, highlights the oppressive potential of the district commissioners' courts. In this extreme case, Bokai was hired as a contract laborer through the government transport office at Sekondi on October 22, 1903. However, after reaching the Tamsu mine, where he had been engaged as the leader of his gang, he allegedly fell ill.[146] In court, he argued that illness had been the true reason for his desertion just days later. He also insisted that he had returned the advances to the manager before he left: "I left Tamsu because I was not well. The manager wrote me about the 10 [shillings] I received from Sekondi. I told him I had given the 10 [shillings] to the [white] headman. I was sick for 3 days. I told the white man I was sick. I went to Lagos Town in Tarkwa. I was sick there. The 10 [shillings] is with the whiteman."[147] The gesture of his reimbursement was not in question, but Bokai was sentenced with "£2, or in default 2-month imprisonment."[148] Current labor laws stipulated that "when a person is imprisoned for non-payment of any money whether as damages, wages or otherwise such money shall be considered as liquidated and discharged, at the expiration of such imprisonment, which imprisonment shall be without hard labour."[149]

Companies in the Gold Coast certainly committed atrocious crimes of de facto indentureship during the twentieth century. Nevertheless, one must keep in mind that penal sanctions in relation to the Master and Servant Act were rare, and that employers used debt only to discipline laborers in a fraction of these cases. Since certain legal and practical obstacles hindered employers from forcibly and systematically keeping African workers in the mines, they instead used other tools to limit the probability of desertion. Perhaps the most effective, if not the most austere, measure of keeping men in the mines was the extended payment system. The majority of workers on agreement in Wassa did not "draw their pay until the final completion of their contract."[150] They did not see any of their wages for the length of their six- to twelve-month contract. They were therefore compelled to rely on a series of small, staggered loans, as mentioned earlier, increasing their potential of falling into debt bondage with moneylenders around the mines. From the companies' perspective, this method of compensation was highly effective, because it got men to stay and continue to be productive in the absence of physical barriers and in the context of inconsistently implemented labor laws. This precise argument is laid out by an administrator in 1903, who commented that "a labourer does not desert when he has some money standing in his credit, except the provocation on part of his employer be unendurable; and there being no monthly payment in his case he does not take two or three days off after pay day to enjoy himself."[151] Some years later the same official insisted that when "a man has a sum of money to his credit he becomes trustworthy."[152] That being said, managers in Wassa also had to consider that laborers became increasingly discontented the longer they lived and worked under such conditions. And as one employer put it, "I do not care that they be bound for longer [than six months], as much trouble arises in consequence."[153] The desperate measures and incoherent strategies of recruiters and employers in Wassa in order to keep laborers on the concessions do not support the argument that individual contract employment "reinforced by many generations of master and servant laws, and supported by different regimes of penal sanctions for breach" was a "powerful weapon intended by the industrializing elites to create an industrial working class out of a subject population which was initially unwilling to go to work for wages" in the case of the Gold Coast Colony.[154] When British officials imagined the future of African workers on the Gold Coast, the notion of class formation rarely, if ever, entered the picture. If there was a latent intention to make some, perhaps temporary, wage workers, for the period under investigation, the colonial stated showed absolutely no urgency about realizing such a goal. Without the assistance of the state, omnipotent capitalists could not exist. The political shortcomings of the mining sector made it easier for African workers to exert power in the face of mine managers' demands.

Workers' Motivations in Their Own Words

There is probably no better way to understand the complex array of driving forces pulling workers away from their rural homes to work on contract in urbanizing parts of the southern Gold Coast than by reading and analyzing their personal correspondences with family and friends during time away from home.[155] Contract workers kept an open line with their home regions, both through direct and regular correspondence from home and through the fragments of news that were passed along indirectly through other migrant workers from the same communities. Loved ones managed to stay informed about their daily highs and lows through these channels. In the letters exhibited here, which can be found in the personal papers of the Chief Transport Officer Frederick William Hugh Migeod, miners shared a great deal of detail about the vicissitudes and opportunities that came with life in these urban centers. In turn, their friends and dependents relayed news about the highlights and troubles of life back home. They provided moral support, reminding the migrants of why they had chosen to leave home in the first place, and assuring them of how their lives would change upon their return. At the same time letter-writing was also self-serving. It was used to inspire regular remittances. Moreover, embedded within the various tales of village life were reminders of the contract men's social obligations, not least to share his wealth with family members. The four letters are presented below in their entirety with a view toward relating their exchanges as accurately as possible. They were written by contract men two at the higher and two at the lower *echelon* of the labor force.

One of these letters was produced on June 22, 1912, in Bonthe District on the coast of Sierra Leone. With the help of a professional letter writer, the otherwise illiterate Conah Mandorrah wrote to her husband, a Kru contract laborer by the name of Gin Bottle, working in Wassa. There were motivational aspects to her note, such as "you went there to find money" and "all people will be looking on you to see what you will bring."[156] She also gave him advice about how to protect his income in the turbulent and largely anonymous mining villages. "You must be careful how you are using your money," she insisted, warning him to curb his gambling habit.[157] Even if the government bank was not widely considered to be an institution for growing one's wealth, among the miners it was significant for its ability to protect cash from others and maybe even from oneself. "If you don't get any place to keep it, put some to the Government Bank so whenever you want it you can take it [from] there," Conah insisted.[158] Not forgetting the dangers that lingered when contract men made their way home, she also warned him to keep his guard up because "robbers" were especially active along those routes routinely taken by contract men moving between Wassa and their rural homes.

In spite of its pleasant undertones, most of the contents of the letter were quite urgent. Both at the beginning and the very end of the letter,

Conah informs her husband that she had recently been sued for twenty-seven shillings. Making things worse, a formal court had sentenced her to live with another man, the chief speaker of Domborkoh, if she failed to come up with the sum by the end of the month. "Send me this amount," she wrote in one instance.[159] "Send that amount quickly," she begged in another.[160] She urged him to borrow the amount if he was in a financial bind, once again emphasizing how easy it was for miners to tap into third-party credit networks. Conah's letter served the purpose of motivating her husband but also of reminding him that "time is very hard now" and that she relied on his support. The fact that he was working in the mines reassured her that he could meet such demands and do so with a degree of spontaneity. Here is the letter in full:

My Dear Husband,
   I acknowledge the receipt of your letter dated the 5th inst. Contents perused with care.
   In reply whereof I am very sorry to hear of the misfortune that happened to you for which you please have my sincere sympathy. I hope you try to be careful of yourself yonder as time is very hard now, and you are the only person [I] am relying on.
   Sorry to inform you that I had a dispute with Soco Beah and the dispute was taken to the barray [barrister], in which they gave me wrong; and the sum of 27 [shillings] was left in the barray. For which dear husband I beg to ask you respectfully to send me this amount. I met the Chief Speaker of Domborkoh and I promised to return this amount at the ending of this present month. If I don't I shall [be] remove[d] from my place to stay with him until you send the money or I get same to give them.
   If no money [is] near you try to loan money and send it.
   In you going I advised you not to play cards any more. I hope you try to keep my advice because you yourself know that you went there to find money and when you come all people will be looking on you to see what you will bring so you must be careful how you are using your money. If you don't get any place to keep it, put some to the Government Bank so whenever you want it you can take it [from] there.
   Whenever you are coming you must please let me know. And in your coming you must be careful as a man who is looking round here told me that you will be deprived by the way coming by robbers. So you must be very cautious or particular. Do try to send that amount quickly. Daddy Baurkie (?) said that you must try to send that amount, and gives you his best compliments.

> Wishing you well
> Kind regards,
> Yours truly,
> Conah Mandorrah, her x mark.[161]

Conah Mandorrah's letter is further fascinating for its ability to crush the image of the passive, slaving, or oppressed African wife left behind in the village or town. Her relationship with her husband was more intimate than the dominant portrayal of husbands and wives in Africa in the existing scholarship. This closeness is imbued in her remarkable degree of frankness.

The remaining three letters were written during the interwar period. The first one is a 1919 letter written to Lacton, a Mende migrant laborer, by one of his siblings and his mother, both of whom lived in Pujehun District in southern Sierra Leone, neighboring Liberia. It paints a clear picture of mining as a well-paid occupation. On the basis of that image, Lacton's family members expected their lives to improve in parallel. The pair expected Lacton to send money back to his home country to pay for both rural customs and consumption. His sister shared that "next April is fixed for the celebration or last burial for grandmas Jani and Basseh and would take place at Meena," imploring "please send in whatever you can."[162] They also were not shy about asking for money for life's simple pleasures. Lacton's mother wanted to "buy tobacco" with the help of his money, either to use for trading or enabling a form of consumption that likely was both personally enjoyable and elevated her social status in the coastal setting. To his relatives, it was evident that Lacton had left home to "make rapid progress."[163] But any progress depended on the spiritual guidance of both ancestors and living family members. Lacton's mother explained that bettering himself was possible only with "my prayers."[164] Finally, as was the case in the previous letter, Lacton's relatives felt it was important to remind him that "the country [is] 5 times harder than when you were here."[165]

> Dear Lacton,
> I am . . . glad to hear that you crawled through the great Influenza. Mama is here not quite 3 feet off me whilst writing this. [She] came purposely to know if [there was] any news from you. I told her you are well. I am ordered by her to inform you that next April is fixed for the celebration or last burial for grandmas Jani and Basseh and would take place at Meena and that you please send in whatever you can. Mama [is] saying: I cannot believe that you are again thinking of

me as a mother and without my prayers you can hardly make rapid progress: the very last of all my issues, and earning money you've not been able to send me even a 1 [shilling] to buy tobacco? Oh no, my boy I was not expecting this from you. Are you not thinking that the country is 5 times harder than when you were here.[166]

The last two letters seem to have concerned contract men of higher professional standing. The first is unique, having been written by the contract man, Alimendi, himself. Alimendi was a contract laborer of either Liberian or Sierra Leonean descent who described himself as working in the government transport department.[167] If he did actually work locally, and not as a subcontractor to the mines, his tasks would have been more clerical in nature. Regardless of his precise professional position, Alimendi appears to have had great success, even wishing to permanently establish himself in Kumase, where his job was located. Although contemporary writings put a great deal of emphasis on the frivolous consumption of migrant workers, Alimendi was evidently strategic with his savings, deploying them in order to achieve an important rite of passage. If migrant workers were perceived as young men when they left their homes to move to distant lands and in order to pursue tough jobs, there was also a decent chance that they would experience social mobility as a result of their stint in the mines. In the letter Alimendi describes having sent home some of his earnings to the local paramount chief, Momo Gbainya, for the purpose of securing the hand of a wife, a young girl just out of the Sande bush school where she was being prepared for married life.

Other points to take from Alimendi's letter are, first, that it was not unheard of for friends to work together having migrated to the southern Gold Coast. Second, we see that his friend Bindi already had at least two wives before he agreed to contract work in Kumase. And "now he has an [Asante] wife," Alimendi reported back home. This may underline a point that has received little attention in the literature, namely that the poorest members of the community were rarely part of these migration networks. Having likely been perceived as unreliable debtors, they would have been discriminated against in the recruitment process for contract work, being forced to stay in the rural areas if they could not fund their own travel. Borrowing money in the context of migration, the creditor-labor agent, who only had a weak personal relationship with the members of his gangs, would have been highly cautious about to whom he gave an advance.[168]

Dear Momo Gbainya,

This letter is to let you know that Bindi and I are working together in the Transport Department at Kumasi and we are both quite well. We are very pleased to hear that you have been made paramount chief of our country. Bindi received a letter saying his Aunt Navo died some time since but now he has an Ashanti wife named Ajira [?]. I want you please to bring Sara out of the Sande bush, and write and tell me when you have done so as I wish to arrange for her to come to me here. I will send money for her. Bindi wants you to beg his brother Josiah to look after his two wives honestly until he comes back. . . .

With all best wishes,
Alimendi[169]

A final trend the letter confirms is a growing pattern of permanent urbanization by contract men. When the letter was written, Alimendi was not strictly interested in settling down in his home region, instead requesting that his wife be sent to him on the Gold Coast. Both he and Bindi chose to maintain their roots in their country of origin but also to establish new ones in Kumase. Many, although not most, workers were doing the same during the interwar period; a healthy economic prognosis in wage work certainly informed their decision.

The final letter, also produced in 1919, was written by the son of a contract laborer, Joseph Bomah, to his father.[170] Joseph was attending the Anglican St. Cyprian's School in Kumase at the time, and his letter is especially illuminating for its demonstration of the generational wealth that some contract men managed to build. Wealth permeates this letter in the form of education and privilege. Throughout, Joseph is insistent about his entitlement to things, whether money, "trouser and coat" for school, or school supplies. He presses his father about the money he promised to send: "You told me at station that if you reached (Tarkwa) you will send me some money" and "you must send me some money about £6." In an attempt to make his father feel guilty, Joseph also wrote, "You brought the money to your wife and your children," insisting that all he needed was just one pound. In his sense of entitlement to money, clothes, and further "things" for Christmas, Joseph shatters the image of the families of migrants being caught in a web of poverty, helping to reproduce wage labor. Evidently, Joseph also knows about English "things," indicating that his father is close to the British "master," and that he had received gifts from England in the past. Even if he did not have the explicit priority of guaranteeing his son a carefree life, doing so must have been part of the father's motivation.

My Dear Father,
  I am just to inform you this few lines, to say that the time when to Tarquah you told me at station that if you reached you will send me some money. You fail to send it. And also I have no trouser and coat. So you must send me some money about £6. Myself I don't think that you will come here again. And you brought the money to your wife and your children. I want £1 from you, and buy me one felt. Try your best as possible. Tell you master that since he went to England I don't saw his face. So, let him try to send me some English things. I am here with your sister Ikaman. I heard that you [are] going to Bongo with your master, for one year and six months. So try your possible best and send me the things, which I tell you on this paper. I have passed my examination. I want all of my Christmas things because you will go before Christmas comes. I have no dresses at Christmas time. I have no more to say again.

I am your loving son,
Joseph Bomah[171]

Recall that in the 1840s, the naval officer Captain William Allen also emphasized that indigenous recruitment was a voluntary (albeit unequal) system. Having been in regular contact with Kru contract laborers, he drew parallels between the gang system and the apprenticeship system that British officials had introduced into Sierra Leone following the abolition of the transatlantic slave trade. Within this system, the moral economy obliged wealthy members of the community to educate or train the less fortunate. Allen communicated that the "commencement of their career is by an apprenticeship to a headman [gang leader]—generally an influential person, and of great previous experience, whose duty it is to initiate his 'boys'—as they are called—into the various duties the white man may require."[172] His use of the word "career" further implies that this was an opportunity with long-term economic prospects. It goes without saying that labor agents benefited financially from this exchange as well, though perhaps not in a dramatically exploitative manner. According to Allen, the members of a gang were willing to make sacrifices for the tools they would receive in the course of their training: "For this preliminary education he [the gang leader] receives a small portion of the wages of each of his party."[173] The letters from African mine workers in Wassa in the early twentieth century testify that the laborers of such gangs were not loyal subjects or oppressed countrymen, but individuals with aspirations (not just needs), who were taking action to participate in the wage market for personal, political, and economic reasons.

The following chapter explores indigenous recruitment and supervision for mining concessions at the turn of the twentieth century, once Liberian labor agents were becoming a rarity, opening up a vacuum for other, new recruiters from other part of West Africa to fill. Indirect recruitment flourished in a variety of forms with the intervention of the government transport department in 1903 and with the increased involvement of more independent individuals who were less reliant on a particular company handing them capital or reaffirming their good character. New edges were beginning to form in the recruitment process for the Wassa mines.

3

Disrupted Recruitment at the Turn of the Twentieth Century

Women, Whites, and Other Labor Agents

The large and flexible supply of Liberians laborers so prevalent in Wassa in the 1870s and 1880s was severely disrupted by the 1890s, when mine managers experienced the rupture of the trusted bonds that they had so carefully built with Liberian labor agents. Instead, new recruitment networks were forged that often entailed fewer checks and balances. The first female labor recruiters appeared on the scene during this period. Even European men now sought to get into this lucrative business. But most of all, African men from other parts of West Africa managed to successfully participate in the recruitment business in growing numbers. The government transport department's recruitment activities for the mines which began around this period will be discussed in detail in the following chapter. This chapter examines the dissolution of the dominance of Liberian labor agents and its aftermath, but especially the opening up of opportunities for new types of labor agents, including some who were ultimately too independent to really be trusted by mine management in the context of what were described as epidemic levels of desertions.

Impacts of Recruitment Bottlenecks in Liberia at the Turn of the Twentieth Century

Two key incidents generated bottlenecks in the recruitment system from Liberia around the turn of the twentieth century. The first was French territorial conquest into southeastern Liberia, starting in the

1890s. The shortage of Kru recruits was immediately felt in the gold-mining concessions of the Gold Coast Colony and its surroundings. In 1896, Edwin Cade, the founder of the Ashanti Mining Corporation, reported to his board in London that although Kru men had been greatly sought-after underground workers until this point, it "is regrettable that the supply is running short."[1] He went on to explain that "the French have put an export duty on them."[2] French officials had first made claims to Liberian territory in 1885, coincidentally, or not, the year that marked the end of the first gold rush in Wassa. In the 1890s, in an attempt to create a colony out of what had previously been the French protectorate of the Ivory Coast, France went on to claim rights to territorial possessions extending continuously from that protectorate into the southern tip of Liberia, beyond the Cavalla River and Cape Palmas through to the town of Garawe, by means of a number of treaties between local leaders in these areas and French naval officers.[3] The French even declared ownership of territory lying in the heartland of the black state Liberia, including Cape Mount, Grand Bassa, and Grand and Little Butu. French officials quickly informed Britain of their intention to establish a formal administration in Garawe, though they would back out of the town on the southeastern coast if they were given immediate control over the area bordering the Cavalla and San Pedro Rivers.[4] The weakness of the Liberian administration in the interior made it impossible to deny these allegations in any definitive manner. This, however, did not stop local officials from adamantly rejecting the agreements and seeking political support from abroad in order to uphold what they regarded as their possession. At the height of the crisis, President Hilary R. Johnson made an impassioned appeal to states in northwestern Europe, but above all to the United States, declaring, "We have no power to prevent this aggression on the part of the French Government: but we know that we have right on our side, and are willing to have our claims to the territory in question examined." Their right, as the president explained, was based on straightforward ownership, but also on their position as Africans, and especially "civilized" Africans, on the continent. This was their "fatherland" after all, consequently their "rightful inheritance." And it was essential that the same rights be preserved for future generations of immigrants of African descent. Perhaps in an attempt to appeal to the proponents of segregation in the United States, he insisted: "We should have room enough, not only for our present population, but also to afford a home for our brethren in exile who may wish to return to their fatherland and help us to build up a negro nationality."[5] The president further asked, "Is there not to be a foot of land in Africa that the African, whether civilized or savage, can call his own?"⁶ He implored the "civilized" Christian nations to allow Liberian officials "free scope for operation" in Liberia for the sake of "fairness" and at very least as a sort of experiment in the race's capacity to self-rule: "Do not wrest our territory from us and hamper us in our operations, and then stigmatize the race with incapacity, because we do not work miracles."⁷ In the end, neither the United States nor any other nation used its influence to arbitrate between the two states. Left with little recourse, on December 8, 1892, the newly appointed Liberian president, Joseph James Cheeseman, finally accepted a treaty drawn up by French officials that granted France territorial rights to the area between the two rivers after depositing twenty-five thousand francs into Liberian state funds. Disputes surrounding the application of the treaty dragged on into the 1920s.

France's 1892 acquisition of the region between the San Pedro and Cavalla Rivers gave it control over a vital source of Kru laborers. This geographical space happened to correspond to the natural boundary of an expansive Kru population during this period. Indeed, Raymond Buell has proposed that "the very fact that this coast was inhabited by a valuable source of labor supply, made France desire its possession."⁸ Although it is not entirely clear to what extent getting access to these laborers was a motivating factor for the whole conflict, as this French map of Liberia from 1835 reveals, politicians there had long been aware of the Kru settlement in the area. French officials did not waste any time before exploiting their new point of access to these men to further their various economic, political, and military agendas around the globe. In addition to levying higher taxes on Kru laborers for export outside the French Empire, thereby raising recruitment costs for foreign employers in many parts of West Africa, French officials continued to mobilize large numbers of Kru men for a variety of jobs on the Panama Canal, in the French colonial army, and in Jamaica, to name a few locations.⁹

A second fateful event occurred just one decade later, when Liberian officials decided to curb labor exports from within the country's own borders. In spite of its losses to France, a significant Kru settlement in Kru town still remained under the jurisdiction of the Liberian government. In 1902, President Garretson Warner Gibson first expressed to his countrymen the government's disapproval of the export trade in laborers from this area.¹⁰ His objection was soon formalized, and by 1903 officials were able to enact legislation severely limiting this trade. From that time forward, only those foreign enterprises in possession

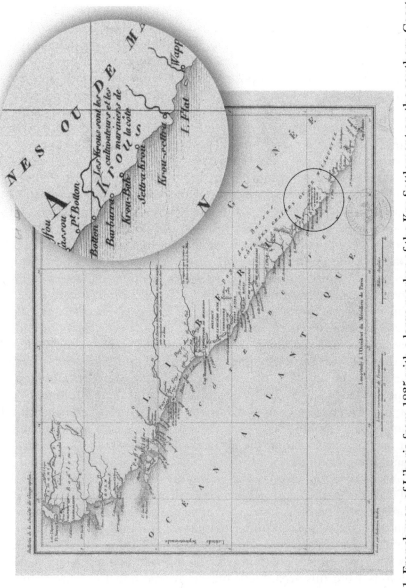

Map 3.1. French map of Liberia from 1835 with a large marker of the Kru Settlement on the southern Coast. Map by J. Ashmun. Bibliothèque nationale de France, département Cartes et plans, GE D-11547.

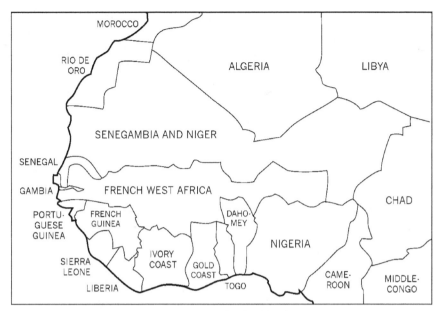

Map 3.2. West Africa, 1902. Map by Talya Lubinsky.

of a recruitment license costing $250 were permitted access to Kru recruits in Liberia. Furthermore, age restrictions were enforced forbidding workers under the age of twenty-one to leave the country entirely. Things were only slightly better for those over twenty-one. Recruiters were now required to pay a fee of five dollars in head money for each of them, as opposed to ten shillings in the past. In addition, regulations were put into place to ensure that the men actually returned home once their contracts were over. To ensure their repatriation, the state held onto a deposit of $150 from the recruiters until each man returned. In consequence, even though the trade from Liberia was not cut off completely, newly enacted laws and regulations immediately put further restrictions on the supply end.

The benefits of controlling Kru labor exports in this manner were threefold: (1) it appeased local planters and land-owning farmers who were desperate for a cheaper supply of hired laborers; (2) it promised to generate greater revenue for the state; and (3) it allowed the government to display political strength in front of its indigenous populations, because the legitimacy of the state was regularly contested by social groups such as the Kru, who lived in the interior and along the coast. The Liberian state was compelled to respond to the petitions of local mid- to large-scale farmers who were facing a time of hardship, because the prices in coffee exports fell by about half between 1894 and 1900 due to their inconsistent preparation methods and the rapid development of the Brazilian coffee trade.[11] Palm oil prices and those of other local agricultural products had also decreased due to more competitive international markets. Rubber was produced only in small quantities at the time. Under such conditions, mid- to large-scale farmers struggled to find laborers who would work on their properties at a wage rate that the planters deemed to be reasonable. They blamed the government for facilitating the exportation of local young men to other parts of West Africa and beyond, locations where they accelerated state-building processes and from which quite a few never returned.[12] As it happened, at the turn of the twentieth century, local officials developed a concrete interest in asserting greater control over Kru Town in the city of Monrovia, and especially over all of the commercial transactions that occurred here, as a means of raising the state's power and revenue. The consequences of these shifts were quite dire for employers, not just in the Wassa mines but all along the coast of West Africa. That same year James Bullen, the secretary of the Mine Managers' Association, requested the assistance of the Colonial Office and the local government transport department for the purpose of labor recruitment "in view of the increasing difficulty of obtaining native labourers, more especially Kru boys and Bassas, owing to the reported increase in head money from 10s to £2."[13] French colonial conquest and shifts in Liberian economic and political policy had a staggered but direct role in transforming the recruitment patterns of the gold mines in Wassa.

In 1902, James Bullen declared that the "difficulty, on account of this deficiency [in Liberian labor], is to find local substitutes."[14] The drastic decrease in the number of Liberian contract men in the employment of the firms around Tarkwa opened the way for gangs from other parts of West Africa. Although members of the association continued to prioritize the establishment of a recruitment monopoly for foreign employers with the endorsement the Colonial Office, in the meantime they met more immediate goals, including casting a broader net for contract laborers from within the Gold Coast and its protectorates as well as other parts of West Africa, such as Nigeria, Sierra Leone, and French colonies.[15] According to the 1903 administrative report, which included a colonial index on the "comparative value and uses of the various tribes usually employed in the mines," mine managers continued to prefer Liberian men for underground mining, because they supposedly "took more readily to drill and hammer work than any other."[16] Nonetheless, concrete signs of a more diverse workplace were also emerging in both underground and surface mining.

Other groups were now being deployed in underground mining in larger numbers.[17] Apollonians (or Nzemans) from the coast, for instance, were now increasingly engaged for this type of work, because Europeans believed that "they have ancient mining traditions dating from Portuguese times."[18] Some Ewe people from the colony and southwest Togo also worked underground, as did the Vais from the southwest coast of Liberia. Surface laborers not only continued to outnumber underground laborers but also diversified during this period. Migrant laborers from other West African colonies were coming to Wassa in growing numbers. According to the same report, "Yorubas and others from Lagos" were hired strictly for these sorts of tasks.[19] Most Ewe people engaged in surface mining, and mine managers also hired Mende and Timmani men from the Sierra Leone protectorate to pursue this type of work.[20] Local men from the Gold Coast were generally only engaged for "desultory surface work."[21] Laborers from the Northern Territories, who later became so critical to mining development after 1906, were described as avoiding gold mining altogether at this time. The Wangara and Moshi peoples were interested in working only as carriers.[22] Nonetheless, Europeans still regarded them as

Table 3.1. Labor in the mines in the Gold Coast and Asante, 1904 (by social origin)

	Origin	Number of men	Percentage
Foreign African laborers	Kru	2,382	13.97
	Bassa	846	4.97
African laborers from other British colonies	Mende	743	4.36
	Yoruba	1,187	6.97
	Hausa	399	2.34
African laborers from the Gold Coast	Asante	2,766	16.22
	Fante	6,545	38.40
	Ewe	1,934	11.35
	Nzeman	242	1.42
Total:		17,044	

Source: Report on the Mining Industry for the Year 1904, p. 5, PRO CO 98/14.

having much more potential; for those who had been hired to work around the mills "settled down to surface work tolerably well."[23] By 1904, of an estimated 17,044 African laborers who were employed in the gold mines in the towns of the Tarkwa and Obuase, less than 20 percent were of Kru or Basa origin, as shown in table 3.1.

If the makeup of the mining labor force in Wassa changed dramatically between the nineteenth and twentieth century, two elements that continued to set this gold rush apart were the reluctant collaboration of state and capital and the direct impact that this political constellation had on the dynamics of indirect labor mobilization for the mines. The introduction of deep level mining after the 1890s demanded even greater control over the workforce, which had to reach a higher degree of efficiency of production. Nevertheless, indirect recruitment in and of itself was not perceived to be an evil. Foreign employers actually still preferred laborers to arrive at the gold mines under the leadership a recruiter who could vouch for them, take care of them, and discipline them during their term of service.[24] On the Rand, for instance, mine managers had learned to tolerate indirect recruitment so long as it was conducted by labor contracting agents who were under their control to a reasonable extent, although early in the twentieth century labor contracting firms were given preference, because they granted both mine management and the government more leeway for shaping recruitment activities. In Wassa, however, the introduction of deep-level mining and professionalization of management was paralleled by the weakening of the managers' control over the labor force. After 1902, mine managers more readily considered entering arrangements with new intermediaries who did not necessarily have a "good book" outlining their reputation and character. As a result, the growing costs of and competition in the recruitment business exacerbated an ongoing problem—desertion.

The Rise of Independent Labor Contractors

The void left by the Liberian "good book" labor agents was quickly filled by a variety of other recruiters and one particular colonial agency, the government transport department. Their presence diminished mine managers' hold on the recruitment process, because the career prospects that the mines offered mattered little to many of the independent labor agents. Now clearer distinctions between those individuals acting strictly as indigenous recruiters or financiers of recruitment and others acting as indigenous supervisors on the concessions (and occasionally the employees of the former group) were also starting to crystallize. More independent labor agents used their own capital to support the mobilization of gangs and also to hire individuals to supervise the men during their term of service. Given the poverty of documentation available for different supervisor-employees, but especially for the growing number of independent recruiters and financiers of recruitment, one can only speculate as to their reasons for seeking out this type of enterprise. Yet evidently, with the splitting of these roles, it was no longer necessary for recruiters to have worked in the mines prior to getting involved in the recruitment business for that sector. Although it was still necessary for the supervisor-employee to be familiar with circumstances in the mining centers in the western protectorate, labor agents or the financial backers of recruitment could come from different backgrounds. Regardless of the extent of their interest in and knowledge of the mining sector, investing money in labor mobilization was a lucrative venture, especially if the right deal was struck with a mine manager. Akan people from the southern Gold Coast, who regularly encountered workers who had recently ended their contracts as well as others who had just arrived from rural areas in search of work, were in a particularly convenient position for this type of business. It is also worth contemplating what percentage of independent labor agents were former moneylenders exploiting the labor power of miners who had fallen into their debt. Getting involved in the recruitment business was an ideal solution both for individuals who personally frowned on wage dependency but appreciated the economic promise of such an investment and others looking to recover precarious investments.

As the case of Quao[25] Hammah from 1894 shows, the trajectory from mine worker to mining recruiter was still quite strong. Hammah lived in Abosso, where he was a petty trader who occasionally worked as a surface miner, "a contractor for firewood for the Abosso Company."[26] For someone who primarily made a living on the fringes of the mining sector through trade in goods, bringing men to the mines was a rather natural progression. In front of court he recalled setting aside twenty-four pounds of his savings, which were intended as advance payments for laborers on the coast.[27] However, he had no intention of personally traveling to the coast. Instead the money was to be handed "to a man Quamina Amman to take to the coast to advance labourers."[28] Amman would perform the actual negotiations for Hammah. Presuming that Amman would recruit twenty-five men, the advance to each individual, including food, transportation, was around twenty shillings per head.

This was a significant amount, but it was also below the thirty shillings (in total roughly thirty-seven pounds, ten shillings for a gang of twenty-five) that companies regularly offered each recruit at the time. Therefore, it is easy to imagine that the men would receive an even larger advance on being hired, and that recruiters like Hammah would already have made a profit through the company's own cash advance system. Even if the recruits were later hired with advances of ten shillings as suggested by the colonial state, twelve pounds ten shillings, or half of Quao Hammah's investment, would immediately have been returned. A certain small percentage would also have gone to Amman for his ongoing services. A closer look at the participation of other independent recruiters in the expansion of the wage-labor market sheds some light on the subject by demonstrating that men like Hammah were not the only ones who funded or conducted recruitment without later taking on the role of supervisor on a given concession. Around the early twentieth century, there were also female recruiters who could not, but more likely did not want to, move to the role of gang leader.

The Hausa Madam Mariam was a prominent and prolific supplier of laborers of all kinds to the colonial state around this period. The gangs she recruited were at first provided strictly for work in government services through the government transport department. A few years later, she was also commended for bringing wage earners to the mechanized mines via the same agency. In 1900, colonial officials celebrated Madam Mariam for recruiting carriers at Cape Coast for the Asante Expedition, the final war between the British imperial government and the Asante kingdom.[29] As noted by F. W. H. Migeod, the chief transport officer, she "collected a very large number during the present war."[30] After the war, between 1903 and 1919, she became a supplier of contract laborers for other tasks, albeit still in collaboration with the transport department.[31] A 1903 letter between Chief Transport Officer F. W. H. Migeod and a member of his staff names her in connection with recruitment of the esteemed Kru labor for the mines: "Please endeavor in some way to get in touch with Lagos men and also Bassa men and Kroomen at Sekondi for supply to the mines. Perhaps that Hausa woman Mariam would be useful in connection with the former."[32] Because the government transport department configured the members of the gangs it subcontracted internally and could find men with experience in the mining sector to command gangs hired in Wassa, Madam Mariam was not expected to do anything else but recruit. In fact, it is quite unlikely that the mining companies ever would have condoned female gang leaders or contract workers in the first place. Until the end of World War I, any permanent female presence in the mining villages was frowned on, so men generally did not consider sending for their wives to join them in Wassa. Nevertheless, the presence of female leaders working in the mechanized mines should not be discounted altogether. Women worked on the mining concessions as part of the surface labor force even during the second gold boom. The existing literature has emphasized that their tasks on the concessions mirrored what women generally did in "traditional" gold mining—panning for gold or collecting firewood. However, it is important to note that although panning might have been the main occupation for women during the nineteenth century, the scope of their work in mechanized mining also evolved with the introduction of advanced machinery. Panning by hand and collecting firewood quickly fell out of fashion or were no longer necessary on many properties by the turn of the century, but women did not simply vanish from the workplace, as a result. Technological advancements called for women to use their skills in new ways. For example, when Decima Moore Guggisberg visited the Abbontiakoon Mines in 1906 she reported on the new roles women had taken on around the new crushing machines.[33] Under a hot tropical sun, she traced the extraction process, starting from when "a big heap of lumps of ore" left the shaft-mouth.[34] "Two natives were very busy shovelling these lumps thickly on to a continuous belt of stiff leather about a foot broad, which was running slowly on pulley-rollers to the crushing-house about fifty yards away, where it made a turn around a revolving wheel and returned direct to the shaft-mouth."[35] Along this traveling belt were stationed "about a dozen women and boys."[36] They were "busily employed in picking some of the lumps off the belt and throwing them away after a cursory glance," singling out the best pieces of ore as in the past. Recognizing that this was a skill that had evolved to meet the challenges of the twentieth century, Guggisberg admired the degree of skill that the women brought to the mechanization process: "These were the pieces which experience had taught them contained no gold, but were merely stones and therefore unworthy of any attention from the scientific machinery through which the remainder of the lumps on the belt were now to pass."[37] She lauded "the unerring accuracy" of these female and male workers. Mine management was equally aware and appreciative of their ability.[38] "Very seldom, the manager assured me, did an unworthy lump pass through his stamps."[39] Official records acknowledge African women workers on mining concessions in the Gold Coast Colony and its protectorate until 1904.

Nevertheless, according to the source materials, as far as recruitment was concerned Madam Mariam was in quite a rare, though perhaps not unique, position. She was described as a particularly enterprising woman and eventually ended up in living Kumase. There she established a station from which to recruit laborers for various types of wage work connected to the construction of various ports and railways, allowing her to accumulate a great deal of wealth.[40] The recruitment business was so lucrative that African labor agents were not the only ones seeking to make a profit from the labor shortage. European entrepreneurs experimented with the recruitment of gangs for the mines during the Wassa's jungle boom.

Although they did not become commonplace, European miners also worked in the Gold Coast and later went into the labor recruitment business. The story of a Mr. L. Eamonson, who in 1902 provided contract laborers for a variety of wage jobs on the mines, as well as porterage, is one case in point. Eamonson originally arrived at the port of Sekondi as an employee of the Ashanti Gold Fields Corporation in Obuase.[41] Soon after his arrival, however, on June 18, 1902, he followed the protocols for voluntarily terminating his contract with the company, explaining that he wanted to get involved in the "transport business" together with G. C. Moor, a chemist and former general superintendent at Obuase.[42] Eamonson's departure was sanctioned, because he had enough credit in his name to cover all of his debts. According to the manager, Eamonson "gave notice to leave our employ two months ago and has paid his own travelling expenses."[43] A manager from the concession reported that Eamonson "is now on his way to the Coast via the Offin River to join Mr. G. C. Moor, late chemist at Obuassi, who I believe is expected to take up this Corporation's transport business."[44] When it became evident that Moor would not actually be returning to West Africa, Eamonson decided to pursue the venture on his own. He initially managed to employ a gang of men whom he later supplied to the Asante mines for surface mining on contract and gathering firewood in particular.[45] He quickly expanded his business by also becoming a contractor for the Côte d'Or concession, where he even constructed a building from which to run his affairs. "His men, as soon as they have acquired sufficient ground, will build their huts on the spot where they are working, so as not to interfere if possible with our own natives," the manager at the Asante mines assured the directors in London.[46] Overall, mining entrepreneurs in the region were enthusiastic about Eamonson's labor experiment. According to one mine manager, "If Mr. Eamonson is able to supply us with fuel at the same rates as we were paying hitherto it will save Mr. James a good deal of worry and labor; and now Mr. Cox has returned and has resumed his old position of bringing in logs to the saw mill etc., has enabled Mr. James to pay more attention to the buildings and other surface operations."[47] Thus, the managers did not expect European contractors to decrease labor expenditures in general as much as they anticipated a more regular and flexible supply of men. Eamonson seems to have been fairly confident in his ability to meet this goal, given the instant luxuries he added to his lifestyle. Right after he had secured the deal with Asante Mines, he began to build a "large house" and to spend "extravagantly."[48]

Nevertheless, tensions eventually arose among Eamonson, mine management, and the African contract laborers. The first problems emerged when the laborers started to demand a monthly payment, which they had come to learn was being offered to other contract men on the same property. Eamonson's plan to keep his African laborers away from the rest of the labor force on the mines had failed. Making things worse for him, mine management openly agreed to the laborers' terms, once again wielding their great bargaining power during this period. As one manager explained, "We did all we could to help him [Eamonson], and as the boys objected to have their money withheld for 6 months (as he arranged on the coast), I agreed to pay them monthly, the same as our other natives."[49] From Eamonson's perspective, however, this was a massive overreach of the manager's authority. Furthermore, the company had made it infinitely more difficult for him to actually keep his men from deserting. The manager, for his part, admitted some responsibility for the fiasco: "I thought that by doing this any cause of discontent might be removed, but evidently our efforts were not appreciated by Mr. Eamonson." In fact, Eamonson revolted, and it was later shared with the secretary of the Ashanti Gold Fields Corporation that he "has cancelled his contract and has left Obuasi, and the opinion formed by Mr. Daw in regard to him has been fully confirmed." As far as Daw was concerned, Eamonson was a "most shiftless and unreliable" individual, who "proved an entire failure as a contractor."[50] It was believed the he had left to provide laborers to the railway, but according to one manager: "I however, think that had he continued with the Corporation in supplying them with firewood, he would have come out all right."[51] In the end, Eamonson disappeared, leaving behind huge debts at the Asante mines, as well as at other foreign businesses. A detailed overview of his financial woes was presented by a miner in Asante: "I may add that he owes the African Association, Cape Coast, £30–9–3, and . . . he is also considerably indebted to this Corporation, viz. £118–4–8."[52] This particular European labor recruiter had big dreams of getting wealthy from the recruitment business in the colony. There is little evidence that many other Europeans followed in his footsteps, as was frequently observed in South Africa.[53] Rather, in West Africa, male African labor recruiters continued to dominate the field.

Labor Recruiters from the Southern Gold Coast and the Poaching Problem

As did other European employers in colonial Africa, mine managers in Wassa in 1902 had to rely on indigenous intermediaries whom they knew to be conflicted between serving their professional appointments and satisfying personal and political ambitions in the accumulation of power, wealth, and honor.[54] Nevertheless, the managers did not aim to establish direct control over individual laborers, either. Rather, the issue at stake was the extent of control they could exert over these middlemen. Indirect recruitment was an accepted part of the gold-mining industry in South Africa. The key difference was the active role of the state in regulating recruitment conditions on the Rand, whereas the colonial state on the Gold Coast continued to approach the topic of sustained intervention with significant hesitancy, as will be shown in the next chapter. In the meantime, the proliferation of labor agents with limited primary interests in the long-term prospects of wage work took its toll. It was not unheard of for agents to cheat or mistreat their recruits in the past. Now, with greater frequency, there were instances of these same agents intentionally pitting companies against each other. Also, with the growing number of (more) independent recruiters from the southern Gold Coast in particular, firms had to deal with the poaching of contract men from their concessions, though it is not entirely clear how widespread this behavior was.

As already mentioned, unlike the Liberian "good book" gang leaders who depended on the authorization and capital of mining companies to mobilize men, many independent labor agents used their own capital to recruit. Consequently they sought to make the greatest profit possible out of this investment, in particular when recruitment was merely their side business or a way to recover a debt. One way in which they sought to maximize their profits was by tricking the members of their own gang. Mine management may have preferred written contracts under the Master and Servant Ordinance of 1893, but prior to 1906 most contracts were made "under hand," with labor agents in charge of the distribution of their men's wages. As was the case on the Rand, they "preferred this system rather than that where the mine paid the workers directly because it enabled them to discount wages in various surreptitious ways."[55] These agents demonstrated great creativity when it came to swindling their men and taking shares of the men's earnings in nonconsensual ways. Some even lengthened their recruits' term of service by continuously deducting additional fees and fines.[56] In 1904, for example, the gang leader John Fry, who worked as a contractor for coal, was exposed for lying to one of his men in order to retain the overwhelming share of the man's overall earnings.[57] The incident came to light after two of the laborers approached the mine's accountant, a U. Taylor. The first laborer explained that one of the white supervisors had promised him fifteen pounds one shilling nine pence for completing his work. Therefore, he was shocked when he received a mere seven pounds instead. He was frank about his suspicions of Fry, blaming him for the unexpected cut in pay: "On payday accused came with £15 and say he give £5 to make house. Accused handed £7 to me. He told me white man only give me £7."[58] When the laborer finally confronted Fry, it was revealed that Fry had indeed received the full amount. Several other laborers confirmed the story, but they struggled and were ultimately unsuccessfully in recovering their full wages from Fry. The gang leader stuck to his story, admitting his crime only when the general manager threatened to get involved in the matter. Fry, a Liberian labor agent whose future job prospects with the company depended on a good reputation, quickly confessed and offered to produce the money, provided the manager "[went] no further with [the] matter." This was the third time Fry tried to "swindle the boys."[59] If Fry's primary alliances lay with an actor outside the mines and if his employment prospects were not necessarily in the mining sector, he would have had even greater scope for deception in his actions against the men he supervised.

Some indigenous supervisors also misused their magisterial powers. The organizational disconnect between general management and regular laborers was particularly palpable in court cases dealing with disputes between gang leaders and laborers. European employers persistently depended on the authority of a indigenous supervisors over the members of his gang and were generally reluctant to get involved in these types of conflicts. The way in which the regulatory system was set up practically obliged European employers to side with the gang leaders. Generally, the Europeans took a stand in conflicts between a gang leader and the members of his gangs only when their own fairness was directly put to question, or when the leader of a gang had clearly broken the code of conduct. For example, when the government contract laborer Shandi was caught quarreling with another laborer by the name of Jusu over "playing cards" at ten in the morning, his supervisor immediately threatened to fine him. Jusu explained that the gang leader said that "they must catch of we and fine we ten shillings each and I say I get no money[;] then the headman [supervisor] say they must [tie] me and they [tied] me about one hour[,] they put me under the sun. One woman come and [rowed] with them before they left me."[60] Shandi refused to pay the fine and even confronted the chief transport officer, who, according to the leader of his gang, had been the one to give the orders.[61]

With the growing separation between the responsibilities of indigenous recruiters and supervisors after the 1890s, recruiting also became more cutthroat. Mine managers repeatedly claimed that desertion was a plague that needed to be dealt with swiftly and severely. However, not all cases of desertion were directly due to laborers' discontentment with work conditions. A growing phenomenon, the poaching of recruits was a crime quite particular to recruiters living in the southern Gold Coast or else ones who had settled there. In trying to better their chances of finding hard-working, motivated men, instead of waiting to encounter some along one of the popular migration routes, recruiters attempted to lure away contract men who were already employed at concessions or in other job sectors in the area. Although this practice may have been common practice among those looking for casual laborers, it was a punishable crime when contract men were involved. As a result, recruiters did their best to organize secret talks with "agreement boys," away from any potential informants to local management. These negotiations were quite similar to those occurring in the home village or at the coast in Sekondi. The labor agents promised to reimburse what their current employers owed them, and cover the private debts that recruits had incurred with local moneylenders, as well as transportation costs. The agents speculated on future wages and even offered a whole new set of advances.[62] Another popular point of negotiation was the array of lifestyle choices available at a given mine. Recruiters promoted areas that had a livelier social scene and more familiar types of "chop," or food, that were missing at the current workplace.[63] Most mines seem to have handed out rice to laborers, rice being a staple food in Liberia, but migrants from other parts of West Africa would have preferred side dishes based on cassava, yams, plantains, maize, millet, or sorghum. For example, an "underground boy" by the name of Albert Kofi from the Tarkwa concession testified about his encounter with an illicit Akan recruiter, stating that the recruiter had walked straight up to him and asked him if he worked. After confirming that he did, Kofi was told that the recruiter "wanted people to work on timber with him and that [we] would get plenty of money to pay our debts."[64] Seven men in total went with the recruiter, Bonsa Kwamin, that day, in spite of having credit in the company's name, because they were told that if they left the mines they "would get plenty of money on timber-work."[65] Each of them was handed a paper allotting them a one-pound advance. But their luck quickly began to turn when Kwamin began to roll back on parts of the agreement: "His clerk told us we would not get a full advance from accused. He refused to pay our train here. We asked him to pay what the mine at Tarquah owed us when we left at his request. We complained to the [general manager] at Abosso and he sent us all to the police. . . . [Speaking directly to Kwamin] We asked you for money at Abosso. £6 was the sum. You would not pay my debt to Adai."[66] Labor agents generally approached contract men while they were away from the leaders of their gangs or out of earshot of white supervisors. They were also known to poach laborers on the very properties where they themselves had been previously employed. This happened when a former policeman, Mana Wangara, of the Abbontiakoon mine, later returned to the mining site to induce contract workers to desert and instead take up work on the Sekondi road. He also offered a local mineworker the sum of two shillings, more than a day's wages at the time, for each additional man he could convince to leave.[67] There were even instances when entire gangs, including the leaders of gangs, absconded. Once a leader could be convinced to switch to a new workplace together with his gang, mine managers were virtually incapable of taking any reprisals against the absconders. Gang leaders were valuable precisely because they kept order and conducted surveillance. Who else could convince the men to stay? Who else could find them when things got out of hands and the men left the property? In effect, managers had little room to retaliate because they relied so deeply on these intermediaries to supervise and discipline. Such an opportunity best presented itself immediately after a gang had arrived in Wassa but before it had reached its assigned place of work. Isaka Dagarti, a labor recruiter and future gang leader for the Tarkwa mine, recalled being approached by another African recruiter at Huni Valley in Tarkwa district as he walked from Osei back to the mine with a freshly recruited gang of thirty men whom he had contracted for six months by verbal agreement.[68] Dagarti complained, "He said I must give him my men to work. . . . There is no food at [Tarquah] Mine. [He] abused me. I said—I brought those people—you can't get [them]."[69] The labor agent was relentless in his approach, ignoring Dagarti and moving to haggle with individual recruits directly.[70] When Dagarti woke up the next morning, all thirty men had left to serve the Offin Rubber Company in Dunkwa.[71]

In another incident, Seiden Wangara, a labor recruiter from the Northern Territories, took the even bolder step of trying to steal several contracted gangs directly from their place of work. Under the pretense of visiting the Dagwin mine in Axim District to collect a debt, he approached a number of labor agents over several days, telling them "there was [work in] firewood at the place from which he came."[72] The leader of a gang at Dagwin, Dogi Wangara, reported his conversation with Seiden to the manager: "He said he wanted to find boys [and] told me to persuade my boys to go away. I said that they were all contract boys [and] if he wanted boys he should go to Coomassie."[73] Wangara was required to pay a five-pound fine for his actions.[74]

The visible demographic change experienced in the Wassa mines after the 1890s was accompanied by critical transformations in the dynamics of indirect recruitment. Cut off from the many agents who had a longstanding rapport with them, and desperate to keep up production, the mine managers were obliged to collaborate with a new group of middlemen. However, the weak loyalty that the growing wave of independent labor recruiters felt toward the mining firms carried tangible consequences for the mining companies, including added competition among individual mines for laborers and greater financial losses due to desertion and poaching, not to mention the extra element of volatility and instability they had to reckon with.

The next chapter discusses another response to the labor crisis. It observes the actions of the government transport department, a colonial agency, that contracted gangs out to the mines. But the agency's story is significant beyond its recruitment efforts. After 1903, the transport department became a leading advocate for the stringent labor policies pushed by the Mine Managers' Association, setting it apart from the larger colonial body politic.

4

Government Strategies for Assisting the Mines

The story of the government transport office complicates the narrative of the Gold Coast colonial government's reluctance to assist the gold mines in Wassa. In 1903, with the approval of officials in Accra, the transport office expanded it duties beyond supplying laborers for government needs and began to assist the recruitment efforts of local mine managers, although the system of labor organization that it introduced was not far removed from what already existed in Wassa. Transport officers subcontracted gangs of laborers out to individual mines for a six-month period against a fee. The greatest change was in the powers invested in the gang leaders, which became severely limited. The leader of a transport department gang did not chose the members of his own gang; he was stripped of all judicial and financial authority; and he earned a wage strictly on the basis of merit and seniority with the agency. At the same time, there were clear signs that Chief Transport Officer F. W. H. Migeod had bigger plans for his bureau. He anticipated a radical transformation in the way of disciplining African workers in the mining sector. To the dismay of top-ranking officials his vision gradually coalesced with those of the members of the Mine Managers' Association. In his determination to support the gold mines, Migeod gradually broke ranks with the larger colonial body.

Proposals for Colonial Assistance by the Mine Managers' Association

With the introduction of deep-level mining, and as a result of the crisis over the shortage of Kru contract laborers, mine managers for the first time came together in a central body, the Mine Managers'

Association, in 1902. The organization rested on the leadership of the Abbontiakoon mines and the Tarkwa and Abosso mines.[1] It was modeled after the Transvaal Chamber of Mines,[2] because most members in Wassa at the time reportedly had "experience in the labour questions of South Africa."[3] It is not surprising, therefore, that paying a fee to contract gangs through the transport department was not what its members initially had in mind when they petitioned for colonial assistance in recruitment. Accepting the assistance of the government transport office actually was a rather controversial choice, because although the office promised to provide the mining sector with a greater number of African workers, the colonial state continued to underestimate the severity of the desertion problem. All in all, the transport office was all carrot and no stick, having little means either to enforce the regular punishment of absconded laborers or to encourage greater cooperation among local firms. Therefore, the transport office's mission of achieving goals such as the lowering of wage rates in the western protectorate had to be taken with a grain of salt. In the end, from the mine managers' perspective, working with the transport department was but a consolation prize. Therefore, even as they yielded to this fleeting solution to the labor question, mine managers continued to petition for more far-reaching and permanent avenues for labor control, including the introduction of pass laws to restrict the movement of workers. To further improve conditions on the supply side of recruitment, they also sought to eliminate all competition by establishing a recruitment monopoly.

One proposal that managers continuously and widely and highly praised was the importation of Asian indentured laborers on a large scale, since they were said to be both cheaper and better-disciplined employees and less likely to get involved in political matters. Percy Coventry Tarbutt was the lead advocate of the "Asian solution." At this time, Tarbutt had already adamantly, and in spite of significant opposition, defended his preference for "coolie" laborers on the Rand, in particular during the reconstruction period after the South African War, when the gold mines faced a serious labor shortage. In 1902, while struggling to recruit a sufficient number of affordable contract laborers in order to increase production levels, Tarbutt insisted on hiring Chinese workers rather than unskilled white miners. It was this outlook that put him at the center of a growing controversy over the influx of immigrant laborers. The scandal came to the fore when Frederick Hugh Page Creswell, the manager of the Village Main Reef mine, of which Tarbutt was a director, leaked to the public a letter Tarbutt had sent to him.[4] This document, which Creswell later read before the Labor Commission of Johannesburg, suggested that managers avoid engaging large numbers of white men for surface work on their property. After consultations with the Consolidated Gold Fields and Wernher, Beit, and Co., managers concluded that "having a large number of white men employed on the Rand in the position of labourers, the same troubles will arise as are now prevalent in the Australian colonies, namely that the combination of the laboring classes will become so strong as to be able to more or less dictate on the question of wages, but also on political questions, by the power of their vote when a representative government is established."[5] Political considerations were often in play when European employers praised the ethics of Asian laborers in the British Empire.

The Chinese labor question also sparked more basic fears about the future ideological and economic path of development of South Africa. On the one hand, the debate put into question the existing racial order within the mining industry and wider economy.[6] The presence of an expanding population of unemployed and poor whites fractured the underpinnings of the ideology of white supremacy. The whites also opposed the importation of Chinese laborers, because even though they worked at much lower wage rates, they purportedly failed to contribute to regional social and economic development to any significant extent. Creswell was known as someone who eagerly employed white former farmers who had lost their livelihood due to the war. According to his bizarre claims, compared with their Chinese counterparts, white labor was more cost effective, because white men had a greater output of labor.[7] In the end, however, he failed to persuade upper management of his convictions.[8] Like the whites already working for these companies, the managers resisted this idea. They were fueled by a desire to continue fostering an image of whites as superior, skilled laborers, as the whites already working for these companies thought of themselves.[9] Tarbutt did all he could to impose this caste-based vision of the mining industry and the broader economy, wherein a small minority of white men would manage a large majority of nonwhite cheap laborers, onto the Gold Coast. In that same year (1902), as the acting director of the Abbontiakoon mines, Tarbutt explained to his shareholders that although it was possible to have African laborers engaged in the simpler jobs that surface mining entailed, the "actual mining will probably have to be done by imported labour to some extent."[10] According to him, this was an unavoidable truth for "those who have the future of this colony at heart."[11] Quite typically for promoters of West African mining at the time, he exaggerated the extent of support that local companies in Wassa could expect from Britain, stating: "I believe that the Colonial Office is devising plans for the encouragement, if not the actual importation of labour from outside."[12] As already mentioned, the experiments undertaken by British officials all fell short of success. Although they gained incremental support from the Colonial Office and the colonial state, scattered efforts of this kind never crystallized into a concrete, steady implementation in the mining sector.

In parallel to their efforts to import laborers on a large scale, Tarbutt and other members of the Mine Managers' Association also petitioned the Colonial Office for support in coordinating recruitment from within West Africa. Even before the founding of the Mine Managers' Association, Percy Tarbutt was one of the most vocal proponents of a recruitment monopoly over labor-dense parts of rural West Africa. He disclosed a plan to get managers more closely involved in the recruitment process on a public stage in 1901: "I have commenced the institution of a labour bureau,"[13] an agency that would eliminate competition for labor among mines. The main idea behind the founding of the bureau, he explained, was "that all the companies requiring labour shall combine and subscribe towards the bureau, thus preventing any indiscriminate touting for labour by those not properly authorised."[14] This was another strategy imported directly from the South African mining industry. As Tarbutt acknowledged, there was "nothing novel in the idea; it has been adopted on the Rand, and has worked admirably."[15] To exercise sufficient pressure on the colonial state, Tarbutt acted "in conjunction with other people very largely interested in this coast," one of whom was almost certainly Alfred Lewis Jones, the Liverpool shipping magnate whose British Cotton Growing Association was an important vehicle for commercial cotton growers on the Gold Coast at the time.[16] Together they presented a proposal to the Colonial Office regarding the establishment of a West African coastal labor bureau to bring men from various regions of West Africa to the coastal regions, where there was a high demand for their labor power. The new labor bureau would recruit gangs in "properly supervised and effectively managed depots" for the purpose of straightening out the recruitment procedure.[17] Tarbutt had full confidence in his plan: "As we develop, no doubt, a proper organisation will be put on foot for the continual supply for labour to these fields."[18] In the end, however, the realization of such a project hinged on the Colonial Office's endorsement. Although Tarbutt hoped "to obtain the recognition of the Colonial Office to the operations of this bureau, the scheme for which has been very carefully worked out, so as not to tread on the toes of our friends at Exeter Hall,"[19] the scheme was politically risky. As far as actors in Exeter Hall, the religious home of the antislavery campaign, and internationally connected groups such as the *Aborigines* Protection Society and its allies in Parliament were concerned, the Gold Coast still had to be safeguarded against forced-labor practices.

About a year later, in 1903, Tarbutt and his associates in the Mine Managers' Association still had not lowered their expectations on the issue of labor recruitment and control. Quite the opposite was the case. The Gold Coast Agency, a subsidiary of Consolidated Gold Fields, sent a new petition to the Colonial Office that year, proposing establishing an agency that, in addition to recruiting laborers in the Gold Coast Colony and protectorate, other parts of West Africa, and China, would also register them individually.[20] This proposal also failed to win favor with Colonial Office officials.

The Challenges of Cooperation

The Mine Managers' Association also attempted to improve conditions in Wassa independent of the colonial state. At one point, its members believed they had the correct tools in place to achieve sector-wide cooperation for the purpose of lowering labor expenditures. Members were introduced to a new wage-fixing policy and no-poaching agenda that the association would enforce through fines. As a case in point, during a meeting on May 11, 1902, the association's secretary James Bullen, who worked on the Abbontiakoon concession, suggested that "mine managers make full enquiries before taking on any gang or boys over [the count of] ten applying for work, more especially Kru boys, and communicate with the manager of the last mine on which they are reported to have been, if there is any reason to believe that they have run away."[21] As Bullen saw it, this was an opportunity for employers to be able to track their men and gain pertinent information about their past performance without any written record. He also anticipated "absconders being allowed to be taken away if sent for."[22] Nevertheless, the full responsibility of inquiring about the employment history of each contract gang rested solely on an already distressed and desperate hiring party. The stipulation was both costly and time consuming, not to mention unworkable in a situation where employers were glad to take on every additional pair of hands available to keep up the rate of production. Ultimately, the system failed, having been built on a framework of self-regulation as it was. Individual mine managers simply did not have enough of an incentive to initiate lengthy investigations into those employees who were willing to work for them, regardless of fears over a gang's or an individual worker's reliability and trustworthiness.[23]

The Mine Managers' Association's vision was not limited to informal communication among various mines either. Members of the association also welcomed additional bureaucracy, as long as it helped control the African labor force. Above all else, they discussed the compulsory registration of individual African workers in Wassa. This procedure would be "effected at Secondi, Axim, Tarkwa or any suitable police station, with a fee of 1s per annum for imported labour [and] 1s per annum for local labour."[24] The manager of the Tarkwa and Abosso mines would also keep a record of the employment status of white laborers "to which all members of the association should have access."[25] And finally, as had been suggested by the governor, Matthew Nathan, in the past, they would appoint a native commissioner, who under ideal circumstances would have "South African, preferably Rhodesian, experience, where the labour problem is more or less similar to that on the Coast."[26] The extant records of the Mine Managers' Association are few and scattered, and what remains is sporadic correspondences between its members and various government agencies. Thus it is difficult to measure the organization's relative achievements. What permeates the records is the group's dysfunctionality. Its members implemented few of the goals it had set for labor control in Wassa.[27]

As already implied, accepting the Colonial Office's advice to collaborate with the government transport office did not distract members of the association from their more radical ambitions. The transport office originally had been set up as a special agency to facilitate and control the recruitment of carriers and messengers for government services and the colonial mines. In 1903, the office began to provide contract laborers for mining tasks, as well. However, the office's involvement in the mining sector quickly turned into something that high-ranking colonial officials had failed to anticipate, as a political synthesis developed between Chief Transport Officer F. W. H. Migeod and the members of the association. The transport office eventually took on the role of mediator between the members of the association in matters concerning their self-regulatory policies. Furthermore, Migeod's private papers reveal the growing collaboration between the Mine Managers' Association and himself for the establishment of a pass law and stricter penal sanctions for laborers in the mining industry.

The Government Transport Department

The transport department was founded in 1901 out of ambitions linked to expansionist developmental imperialism. With backing from the Colonial Office, it and a few other recently formed departments and worked under the assumption that the mining sector actually had at least some potential to be a vehicle for development. At the time, a new colonial state with a new organizational structure had begun to take shape, and the transport office was one of many newly established agencies with specialized functions. There were several clear signs that the colonial state was reevaluating its relationship with the mines. In 1900, Governor Matthew Nathan personally visited Tarkwa in order to meet with the leading men of the mining industry and get a clearer understanding of current labor conditions and the working of the Concession Ordinance of 1900.[28] Between 1901 and 1902, at a time when modern transportation structures connecting the coast and interior of the Gold Coast were either absent or under construction, transport officers arranged the hiring of carriers and messengers for government activities and the mines. When it came to tasks related to communication, conquest, and straight-out war, the transport officers were responsible for gathering the necessary manpower. In 1902, the department managed a single station at Cape Coast for the recruitment of carriers.[29] By the following year, however, its operations had expanded to include a total of five stations in various protectorates of the Gold Coast, including the towns of Dunkwa, Obuase, Kumase, the Cape Coast, and Sekondi, each being staffed by one European official and a few African clerks.[30]

In 1903, Sekondi was designated as the principal station.[31] Relocating the transport office's headquarters to Sekondi signified a deepening of relations between the chief officer, F. W. H. Migeod, and the mining entrepreneurs. This port city was a crucial administrative point through which all mine managers passed. Having its headquarters there also facilitated recruitment, because it was "doubtful whether an agency at Cape Coast had any access to the class of labour required on mines."[32] Liberian laborers had always traveled through Sekondi to reach Wassa, and even after the 1890s, people came here in search of wage work. Therefore, an administrator who contended that the presence of the transport office played a key role in attracting men to the city was likely misguided: "Before the transport department moved its headquarters from Cape Coast Castle to Sekondi there was virtually no free labour obtainable at the latter place, and there would be very little again were its headquarters removed elsewhere."[33] The transport office certainly attracted more men, but it is important to keep in mind that the number of independent indigenous labor agents scouting the area also was on the rise after 1902. The government transport department was mostly significant because for the first time ever, it created a pool of wage laborers to which foreign mine managers could gain access at will and without the fuss of having to negotiate with African middlemen, though this pool was certainly limited in size.

Governor Matthew Nathan wrote to the Mine Managers' Association in 1903, stating that he "regretted that use had not been made of the native labour and transport agency at the Cape Coast. . . . The government transport officer could adapt his office to the supply of mining boys and native craftsmen."[34] Not too long after, some of the most successful mining companies in Wassa—including Cinnamon Bippo Mines, managed by Thomas Birch Freeman Sam and William Edward Sam, Jr., and the Abosso mines, managed by Gerhard Stockfeld—started to utilize its recruitment services, albeit with some reluctance.[35] In terms of impact, the chief officer reported that his department had provided a "considerable" though "not great intrinsically" number of laborers to the mines that year.[36] The number of laborers, including carriers, who had been contracted to the mines that year amounted to 2,049.[37] During the following year the officer noted that "the mines have not had to apply to any great extent to the transport department for labour." Nonetheless, contributing to these declining numbers was the fact that "when any labourers have been supplied and have completed their engagement and been paid off, they will frequently go back to the mine on their own account."[38] In 1905, transport officers sent exactly 300 contract men to the mines.[39] This count dropped to 216 in 1907.[40]

The transport office also added further regulations to the recruitment process for European employers, and to some extent, for the laborers themselves. Recruitment through the department was probably the most straightforward and least risky manner of hiring a gang for contract work after 1903. When a company needed additional men, mine managers made requisitions directly to the department, forwarding a brief telegram notice in which they specified the number of gangs required and the length of service for which they were needed.[41] The communications relating to such orders were technical, short, and to the point. For instance, in one of his letters, Chief Officer Migeod passed on a request to his assistant transport officer at Obuase for "two gangs to work at

Tamsoo Mine Tarkwa for six months or a year." He advised his assistant to approach the indigenous supervisors Seymour and Johnson for the job, if they were willing. With their passbooks in hand, they would be asked to head to Effuenta, leading a group of twenty-four men each, ideally of the Mende people. Their subsistence being set at three pence daily, the assistant transport officer now had the responsibility for weeding out "any undesirable men."[42] Directions to the Tamsu Mine were anything but precise, perhaps an indicator that the gang leaders already knew their way around the area. Migeod wrote the following in his request: "Address the gangs to E. V. Benusan [Esquire] at Effuenta. His house at Effuenta is on a hill lying between the Government Hill and the railway line, and about a mile below Tarkwa station." Most likely the assistant transport officer was the only one who needed the additional clarification that, "Tamsoo is a little South of Effuenta."[43] Looked at from this perspective, hiring through the transport office presented a simple and straightforward option for the employers, who even enjoyed a degree of quality control as a result.

At the same time, the new securities and general improvements that the transport office provided were embedded in a web of bureaucracy, fees and regulations, so that it failed to ameliorate the problem of high labor expenditures. After paying a variety of deposits and fees, mine managers had to deal with what many regarded as unnecessary and unprecedented interference, as transport officials aimed to ensure the fair treatment of their gangs while making their own services sustainable. The daily wages of laborers engaged through the department was a low one shilling per head. Advance payments were an uncompetitive ten shillings, intended solely to entice individual recruits. Also, according to a departmental report, "The customary advances of ten shillings made to each labourer on enlistment by the transport department is refunded by the employer and charged in the labourers' first month's pay sheet."[44] This sum was refunded in the form of a deposit of around twenty pounds per gang. The *Government Gazette* reported that the transport office held on to this fee, using it to resolve any discrepancies in pay. At the end of each month, pay sheets were sent to the transport office at Sekondi with a check for the amount due to the gang, and the deposit was "held to draw from in case the transport office considers fines, etc. exorbitant, etc."[45] Although the transport office absorbed a share of the deposit, it was only partially liable to the mining companies. For example, as outlined by the *Government Gazette*, if a laborer "deserts before the advance is worked off the employer bears the loss of such part of the advance as is not worked off."[46] The agency did at least provide substitute laborers in certain instances of abscondment. As long as "ill-treatment" was not the driving factor behind a laborer's leaving without notice, the transport office guaranteed that another laborer would complete the contract for "the period the original man was engaged for."[47] Overall, hiring procedures through the colonial agency were quicker and more clear cut when compared with other avenues. Moreover, at least some of the costs associated with desertion could be recovered.

Administrative expansion through the transport office also entailed a greater interest in the welfare of African workers. In trying to gain control of the labor market, the department institutionalized certain practices to improve the lifestyle and productivity of their staff by means of greater health care and by limiting the potential abuses of gang leaders. The gang leaders were stripped of their authority and made heavily dependent on the department in all aspects.

In 1903 the department published its *Rules as to the Permanent Carriers in Ashanti and the Northern Territories, and Laborers etc. supplied by the Transport Department to the Mines.* One point of contention between employers and employees that had to be clarified in the government contracts, was when laborers had the right to demand food. The rules stipulated that the subsistence of gangs hired through the government "must on no account be stopped whether for punishment or any other reason":[48]

> Several amendments have been made in the terms on which labourers were engaged, especially with regard to pay for days on which they do not work. The idea that they should draw their full monthly pay minus deductions for absence without leave, the mine managers said, did not work well. They preferred simply giving a man credit for the days he casually worked. The chief question then hinged on Sundays—whether the employer or employee should bear the cost of the day off. At a fixed sum per month the loss fell on the employer, although nominally on these terms the whole time of the labourer was at the Manager's disposal.[49]

As this quote demonstrates, the provision of food on nonwork days was a particular point of contention between the transport officers and the mine managers. From Chief Officer Migeod's point of view, this issue was especially dire, because some men had neither the money nor the necessary social network to get them through any form of hardship. He explained, "I was obliged to make it a hard and fast rule that the labourers must receive their subsistence whether they worked or not. In the case of men drawing their pay monthly, this would be an unimportant detail; but it is far from being so with those who do not draw and pay till their contract is finished."[50]

The government transport office took further steps to deal directly with individual workers and to minimize political aspects of the relationships between gang leaders and laborers. To minimize abuses, gang leaders employed by the colonial agency were not in charge of hiring individual laborers, nor did they choose the members of their gangs.[51] They also did not distribute wages to the members of their gangs had earned, for example by distributing their wages from a lump sum.[52] In 1903, the government transport office announced that all laborers supplied by the department would "be paid off by the transport officer at Sekondi,"[53] and that companies therefore had to provide subsistence until they had reached that station. In other words, even at the end of their contracts, the government gangs also had the right to certain amounts of food and drink. The contract stipulated that companies provide subsistence at a rate no less than that which would cover a worker's passage to the government transport office where he picked up his wages: "On discharge, a labourer or carrier must always be given enough subsistence to reach Sekondi. If he is sick, and likely to be a considerable time on the road the subsistence money must be increased accordingly."[54] It was up to the laborers to distribute the funds evenly throughout the trip. Once the term of service was over, the workers traveled to the coast to pick up their wages from the transport office, where they had been "deposited with the transport officer monthly by whichever mining company had temporarily engaged them."[55] Each man could claim his pay directly from colonial authorities, approving or appealing individual deductions.[56]

In addition, the gang leaders working for the department were wage dependent; their wages varied "with their experience," as seen in table 4.1. Also, their wage rates were initially not much higher than those of unskilled laborers, though they could improve with good behavior and seniority. Overall, the institutionalization of the gang system had the effect of weakening the transport office gang leader's authority.

Although the government transport office's new security measures were extensive, they were not all encompassing, and conflict continued to ensure between gang leaders and the members of their gangs. The small loans given to laborers by their leaders still were a source of tension. For instance, this was the case when the African supervisor Johnny loaned the laborer Kiawo six shillings, and Kiawo kept postponing a full repayment: "Kiawo owed him six shilling and paid him two

130   Chapter Four

Table 4.1. Pay of gang leaders hired by the Government Transport Office (varying by experience), 1903

	Pay	Subsistence
On first appointment	1s	6d
After six months, if satisfactory	1s 3d	6d
After one year	1s 6d	6d
Captains	2s	6d

Source: Transport Circular No. 3, Official Papers on Labour Policy, July 11, 1903, CUL RCMS 139/12/10.

shillings six pence leaving a balance of three shillings six pence."[57] The laborer later concocted the story of another laborer, Lasana of Baioo Gang, who supposedly owed him some money.[58] Nevertheless, when Kiawo later was confronted in front of the Baioo gang it turned out that Lasana was not in Kiawo's debt at all.[59] Johnny ordered Kiawo "to bring anything he can get to put [up] for pawn[ing]."[60] Kiawo handed him the cloth (whose value was intended to cover the debt) in front of two other members of the Baioo gang, named Blackie and Dirty Water, though he later reported to the transport office that his gang leader had stolen it.[61] A second case indicates that gang leaders working for the transport department still had the means to compel individual laborers to stay on a particular property beyond the length of their contract. The laborer Karthar complained that his gang leader had taken and refused to return the amount of one pound ten shillings from him. As Karthar explained, this had simply been a ploy to extend his service at the Nsawam concession.[62] Although the incorporation of this system of labor organization into the colonial administration did not preclude abuses by gang leaders altogether, the transport office made great efforts to curb these types of activities on a case-by-case basis. Besides endeavoring to increase security in the wage-labor market, the transport office also worked harder to eliminate competition and foster cooperation among the mines in order to reduce average labor expenditures.

The Government Transport Office as Mediator

Transport officers assumed a mediatory role not only between gang leaders and laborers but also between individual companies in and around Tarkwa, Tamsu, and Abosso. It has already been mentioned that given persistent shortfalls in the supply of labor, mine management had developed the destructive habit of offering cash advances haphazardly. To counteract this expenditure, transport officers had encouraged lowering wages on the condition that African miners received three pence subsistence daily. Yet, even with Chief Officer F. W. H. Migeod as mediator, extensive cooperation between the mines remained elusive. Some mine managers worried that the plans of action dictated by the Mine Managers' Association would not be adopted by all companies, either to an equal extent or at the same time, which would put those firms that did adapt early at a grave disadvantage. To illustrate this point by way of a specific case, in 1903, J. Fletcher Toomer, the general manager of the Prestea mining company,[63] wrote to Migeod expressing his concerns regarding a proposal to lower the wages of the African workers. He apologized: "Although I informed you, that it was my intention, to lower the wages to the scale agreed upon by the Mine Managers' Association, I have not yet done so as there seems to be considerable uncertainty amongst the members of the Association and until the reduction is unanimously agreed upon and date fixed for it coming into force, I cannot be the one to start it, as it would result in my losing all my best men."[64] Toomer was managing one of the most successful companies at the time. He was also a registered member of the Mine Managers' Association. However, because of his concerns, he continued to distribute wages higher than those agreed on by the organization. In an early phase, Migeod attempted to combat these colossal challenges with lone-wolf tactics.

The Wassa mines were already in a disadvantaged position compared with gold mines in the Asante region to the north. In contrast to the Asante mines, the Wassa mines were located within a geographically limited and tight urbanizing space that was conducive to job hopping. However, Toomer's conundrum illuminates a further reason that cooperation remained elusive, namely that different concessions had varying degrees of access to particular resources. Distrust among the firms compounded with structural inequalities, making mining recruitment an inherently uneven quest for the various mining companies involved. Generally speaking, the introduction of the railway in 1903 was a great help to the Wassa mines. However, not all companies benefited from its presence to an equal degree. Tracks ran through certain properties but not others, affecting a concession's access to goods and people. Those properties that were not located directly along the rail lines faced a different set of fears when it came to the implementation of capital-friendly regulatory measures that the Mine Managers' Association had decided on but that nevertheless threatened to affect them disproportionately. Toomer's Prestea concession was one of those properties with no immediate physical connection to the railway until 1908.

Above all, competitive recruiting was such a serious issue for employers in Wassa because it fueled desertion rates. Letters preserved as part of Migeod's private papers—some of which were written by workers and addressed to Migeod personally, others produced by mine managers in response to the allegations of laborers—show that few other remedies were available for this evil. Although certain legal penalties for absconders were quite harsh, most of these individuals eluded the colonial judicial system—the same system that failed to protect them when they were discontented with work conditions.[65] Workers were generally likely to become litigious only if they still had outstanding credit with the company against which they were making allegations.

Migeod regularly struggled to mitigate conflicts between gangs and employers in the mines. He employed a strategy of "soft" intervention with those men hired or hiring through his agency. For example, in 1906, a gang contracted to the Offin River Company in Dunkwa wrote to Migeod to complain about a hot-tempered white foreman who was troubling them: He was "always beating" them and "always get fever" in spite of their claims of having exemplified a good work ethic.[66] The gang also reported abuses by a particular African clerk working on the property, accusing him of "not keeping our time properly." He even threatened to deny the men their wages entirely, since it was in his power to report that the men "never do no work."[67] In their correspondence, the men appealed to Migeod's self-understanding as an overseer of such labor disputes: "So we are obliged to let you know all this, as to write our big master or the captain here, to less[en] our trouble."[68] The gang had already been working for the company for at least a month and had accumulated credit with the company.

A similar pattern of protective action by the transport office had occurred in 1903 when, following a confrontation with his employer, a gang leader by the name of John Aggrey traveled all the way from the mines back to Sekondi to give testimony about his gang's troubles at work. Aggrey presented two complaints. First, he accused the newly hired general manager of the Tamsu mine of making excessive demands of the laborers. Migeod agreed that the evidence supported the laborers' story. In his remonstration of the manager, he wrote, "Judging from the pay sheets I already have, they have been worked sufficiently hard; and I should not care to see in the rent pay sheets—those for October—that they had been unable to keep to anything approaching fulltime."⁶⁹ Aggrey's second and main grievance lay in the fact that he and the other laborers were not going to receive their hard-earned wages. His gang was being fined for refusing to work underground, even though this work had not been stipulated in their contracts. Migeod referred the manager to Transport Circular No. 3, from the *Government Gazette* of December 31, 1903, which read: "Labourers employed underground on the mines to receive 3d more a day extra pay" and that such work was "not to be compulsory."⁷⁰ Migeod felt compelled to remind the mine managers that the workers "are surface men, and underground work is entirely optional to them. Should they undertake it, they receive 3d a day extra pay."⁷¹ The manager's behavior was clearly in breach of contract. With his action, Migeod shone a new light on the regulatory potential of the colonial state. Arguably, however, it was not so much his presence that convinced men to ask for mitigation procedures, instead of just deserting a company. Rather, laborers chose to take advantage of the agency's protection only in order not to lose their earnings.

Naturally, other issues trumped the laborers' moneymaking ambitions. In 1903, a group of African carriers sought the assistance of the transport officers when their employer failed to supply them with rations of food on Sunday and denied them any credit. The leader of the gang took their complaint to the transport office by way of a messenger, forcing Chief Officer Migeod to send a warning to the accused manager in East Akim: "Sir, A carrier has been sent to me by the headman [gang leader] working for you to say that they do not receive subsistence for every day in the month. As this is compulsory whether they work or not, I shall be obliged if you will issue the necessary directions that it be not stopped again. If any of the carriers require an advance will you please also let them have it, changing same in pay sheet."⁷²

In other instances it was not monetary gain but grave maltreatment that led gangs to initiate inquiries with the transport office. In 1903, Migeod had to reprimand managers at the Anfargah Mining Company following multiple accounts of corporal abuse. When five government subcontract laborers were "struck with some weapons on the mines and wounded," Migeod supported their refusal to return to work. Although a strikethrough applied to some of the damning accusations in the original source shows that Migeod ultimately wanted to show the manager more leniency than he actually deserved, he explain that at the very least the men would be compensated for their labor: "I am informed that there are some other labourers seriously injured at Tarkwa. These will also receive their pay when they arrive.⁷³ I will let you know later of the remaining absentees as I am not finished with all of them yet."[74] The case against the manager was compelling, because other African workers had had similar experiences on that concession. In August 1903, tensions between laborers and management came to a climax at the Anfargah mines in Tarkwa after another bout of extreme violence and common disregard for cultural sensitivity. Contract men sent there by the government transport office left the property without any notice. The details of the incident leading to this strike were outlined in Migeod's letter to the mine manager, Philip Poore, Esq.: "Sir, I beg to inform you that a letter has been forwarded to my office signed by some of the headmen [gang leaders] at your mine, in which they complain of excessive tasks, and also more particularly of being beaten with a stick, and being prevented from searching for the body of a man who drowned himself."[75] Migeod recognized that the manager was "having some trouble with the labourers,"[76] but in his correspondence, Migeod did his best to remind, or perhaps convince, the manager that treating the laborers well was in the company's self-interest: "I however, hope by now things have quieted down; but I trust you will inquire into the men's complaints, and adjust them so far as they are reasonable, for should by any chance Anfargah acquire a bad name, whether justifiably or not amongst the natives, it will be practically impossible to get any to go there in future."[77]

These uncomfortable and up-close insights into labor relations, combined with having only a minimal amount of power at his disposal to change them, ignited in this colonial agent a deep desire to obtain a fixed set of strategies with which to tackle some of the problems the mining sector was facing. For although Migeod was invested in the health and productivity of African laborers, his loyalties most clearly lay with the employers themselves. Following the conflict with J. Fletcher Toomer of the Prestea mines, Migeod petitioned the colonial secretary for an extension of colonial law in Wassa: "I understood it was generally agreed that the members of the [Mine Managers' Association] would not compete against each other, but when I saw Mr. Toomer here, he said if he wanted labourers quickly he would go about it his own way, and he distinctively refused to recruit on even terms so long as he wanted men."[78] He described Toomer's action as "being detrimental to the public good, and also an obstacle to my endeavor to reduce wages for the coming year."[79] Migeod hoped that "legislation in the subject may be hastened."[80] Until then, the actions of a single manager such as Toomer could offset all mediatory efforts made by the transport office. In Migeod the Mine Managers' Association had finally found an ally

Government Strategies for Assisting the Mines 135

within the ranks of the administration to help them implement more extensive and widespread labor regulations.

Chief Officer F. W. H. Migeod: A Colonial "Man on the Spot"

According to Lawrance, Osborn, and Roberts,[81] during the formative years of colonial rule it was often European "men on the spot" who developed "colonial policies and procedures in response to local circumstances."[82] "The "rules" of colonialism had not yet been developed or established. Therefore these men had to, and did, practice ad hoc colonialism, usually with the help of loyal African intermediaries in what was still a very malleable colonial apparatus.[83] Men in the field were deeply involved with issues that at first seemed of little immediate concern to officials in centers such as Accra. In 1903, Migeod found himself in a position of great authority due to his physical distance from the political hub of the colony, a position in which he could have a lasting impact on policies surrounding labor relations and control on the Gold Coast.

With his background and formal upbringing, Frederick William Hugh Migeod sits comfortably among the ranks of colonial administrators of the time.[84] He was born on August 9, 1872, near the rapidly developing rural parish of Chislehurst, near Folkstone, Kent, where he was privately educated.[85] He gained some military experience, but unlike those of many higher-ranking colonial agents in Accra, his résumé did not include a series of expeditions followed by prestigious promotions. In 1889, he entered the Royal Navy Pay Department as a clerk and four years later was promoted to the respectable position of assistant paymaster.[86] In May 1898, Migeod resigned from the navy to take up his first Africa assignment, in the West African transportation services. By September 1, 1898, he was engaged as assistant transport officer in northern Nigeria for the West African Frontier Force (WAFF); less than a year later came his appointment as supervisor of customs in the Gold Coast Colony. A position as assistant transport officer for the Ashanti Field Force followed a few months later. Migeod then worked as supervisor of customs in Accra.[87] Finally, in September 1901 he accepted a position as chief officer of the Gold Coast transport office, which he occupied for almost two decades.

Migeod, like many other colonial agents, missionaries, or laymen residing in remote parts of Africa, quickly became engrossed with his new surroundings, taking advantage of his regular contact with Africans to explore and enhance a multitude of interests in language, culture, and science. During this time, he contributed to a number of academic fields, including anthropology, linguistics, mathematics, plant science, and history, not unusual for this period. He even authored a few books, including *The Mende Language (1908), Languages of West Africa* (2 vols., 1911), *Hausa Grammar* (1914) among others. A number of his articles were published in the Royal Anthropological Society's *Man* journal: "Antiquity of Man in West Africa," "A Mende Dance," "The Racial Elements Concerned in the First Siege of Troy," and "Supposed Duodecimal System in Burum Language."[88] Later in his career, Migeod also became a member of several academic societies dedicated to the topics of culture and society, as well as a Fellow of the Royal Anthropological Institute, the African Society, and the Folklore Society. Thanks to a chance encounter with the American paleontologist George Gaylord Simpson, we witness a harsh caricature of Migeod as "a dried up fellah that looks as if he might have a fairly sticky past tucked away somewhere."[89] In a comical letter written to a close confidant, Simpson chastised Migeod for being an Africaphile, who was likely used to

> week-ends at Brighton (where all the beautiful and damned go with the not-even-handsome but equally-damned) or that sort of thing—but otherwise neither here nor there, as they say in Nigeria. Well, oddly enough Nigeria is to the point, for I will say that he knows his Benin as well as the Obi himself does. This little prune has tramped right across Africa on the equator ([and] that's a fair jaunt, mind you) [and] then not yet sufficiently fed up he tramped right back on the parallel 5° South, which is even worse. He thinks Lake Tchad is the height of civilization and he's made his tea out of water from the various sources of the Nile. The upper Congo rather bores him, but the Gold Coast is amusin' because there's some jolly good snipe-huntin' theyah [*sic*]. He's never had any trouble with lions or leopards—oh! Of course he's had to kill one now & then that came bothering him, like you would a mosquito—but the fellows that get messed by those vermin go lookin' for them, and he hasn't lost any lions. Just what he has lost that he goes wandering through tropical Africa and can't wait to get back he doesn't know, himself.[90]

To return to his activities on the Gold Coast: Migeod's power, that of the government transport office, lay precisely in his autonomy and distance from Government House in Accra. The duty of the transport office had to be considered under two heads: "financially, and from its practical utility to the colony both from a mining point of view and from a Government point of view."[91] Yet regardless of its contributions to the colonial state, the agency could operate only so long as it raised its own finances.[92] Migeod balanced his budgets with due diligence but still drew negative attention from high-ranking colonial authorities when his department pushed for a greater and more effective administrative presence in Wassa, for the purpose of labor control. The pinnacle of the partnership between the government transport office and the Mine Managers' Association was their lobbying effort, starting in 1903, to pass the Concessions Labor Ordinance, which called for individually registering all laborers in the gold mines, a policy that had been endorsed in South Africa in 1895.[93]

The Concessions Labor Ordinance of 1903

In contrast to those in East Africa or South Africa, colonial governments in British West Africa did not rigorously implement labor laws in a way that would systematically benefit European employers by forcibly keeping men in the labor force for extended periods of time.[94] Though never enacted as law, the Gold Coast's Concessions Labor Bill of 1903 presented a distinct move in the same direction. It devised a strict and structured alternative to what many foreign employers perceived as the lax conditions of labor control in the Gold Coast Colony. Amendments to the Master and Servant Ordinance in 1893 lobbied for by mine managers did nothing to encourage more consistent enforcement of labor laws, though they did subject those laborers who were caught after deserting to harsher penalties. Both the Mine Managers' Association and Migeod imagined that the proposed bill would curb desertion and improve retention rates for the companies by restricting the movement of laborers. They complained that "they have absolutely no control over their men who run away from one company to another and this entails heavy expense."[95] If a laborer had absconded, it was often difficult to identify the offender, because generally identification was possible only with the cooperation of a compliant gang leader.[96] When the leader was in favor of absconding, however, there was no way to "prevent whole gangs of boys leaving one mine to go home or obtaining work at another after having promised to work a definite period."[97] Out of this frustration arose an appeal to the chief justice of the colony in 1903 for legislation to give employers tighter controls on the labor force. In an effort to keep workers from moving between mines in a disorderly manner, the Mine Managers' Association argued that "if, were they obliged to bear an official pass as being in search of work or one releasing them from their past engagement [and] were also obliged to obtain another, at or through the mine of engagement or reengagement, that evil would be practically obviated [and] the chance of dodging the regulation would be reduced to a minimum, more especially if even a moderate number of reliable civil or mine police were vested with authority to demand the passes of natives in the open street."[98] Moreover, it would mean "the registration of all mining and floating populations and need not at present apply further and I see no reason why it should be either a very impossible one to work or even expensive, since it should be self-supporting."[99] Mining entrepreneurs had already petitioned the Colonial Office for the implementation of pass laws prior to 1903.

That same year the transport office initiated a registration and identification program for its own laborers, "which was only applied to those persons who sign on for definite periods."[100] When the transport office enlisted contract men, officers proceeded to take prints of all ten fingers. In contrast to standard practice in the British Isles, these fingerprints were not merely identified with the worker's name but certain tribal markers were recorded, reflecting an understanding of African tribal society. The transport office suggested, "As against the system established in England it must be borne in mind that there is a natural division of the population here into tribes who are unmistakably distinct."[101] A counterfoil of such a certificate kept in Migeod's records lists a number of other ways to identify the worker. In language particular to that period, the form asked for country name; other name; sex; age or apparent age; race or tribe; trade or profession; personal characteristics; height; color; tribal marks; other marks; whether of stout or slender build; and it required a signature of those workers who were literate.[102] The thumb mark was of especial interest to the officer: "Every individual item may be inexactly entered owning [to] carelessness or other causes, but the thumb mark, if clear, cannot fail and in only one case out of 3,000 have I mistaken a man through similarity of thumb mark owning to a defective impression, and a further scrutiny of the ridges revealed difference."[103] This pilot project, which appeared to be no more than a steppingstone for the determined Migeod, suggested a colonywide registration scheme.[104]

The scale of Migeod's ambitions in these matters put him at great odds with officials in Accra. For example, in July 1903, just one month after the Mine Managers' Association had appealed to Governor Matthew Nathan for approval of a pass law, Migeod submitted a Transport

Circular to be reprinted in the *Government Gazette*. It included the section "Regulations for Registration of Labour," which would have become law had the circular been published. Under these regulations not only the department's own contract men would be registered. Instead, according to a draft of the circular held in Migeod's private papers, the proposed rule pertained to "black and coloured males and females not belonging to any of the tribes inhabiting the Gold Coast colony or hinterland who come to work in the colony as carriers, cooks, artisans, etc."[105]—in other words the majority of the contract laborers in the colony and its protectorates. If the draft had been approved, these laborers would have been required to fill out a certificate at one of the transport offices to be handed over to their employer on getting hired.[106] Laborers would be identified and tracked by means of their certificates.

Nevertheless, even though the government transport office promised to take on the job of issuing these documents, Migeod was not successful in winning support for his plans from the higher echelons of the administration. Governor Nathan signed the circular for the *Government Gazette* of August 1903 only after this proposal had been taken out.[107] This was not a surprise, because the governor had already been under significant political pressure from the Aborigines' Rights Protection Society, which opposed many of the mining entrepreneurs' demands. However, this was not a total loss for Migeod and the mine managers. Their bold move seems to have been a catalyst for the governor's serious consideration of a formal bill that went in a similar direction. The Concessions Labor Bill (1903) was drafted and put before the governor not long thereafter with the consultation of the chief justice.

William B. Griffith, who served as chief justice of the Gold Coast from 1895 to 1911, might have been one of the fairest reviewers possible in the higher ranks of the Gold Coast government. Griffith was the eldest son of Governor W. Brandford Griffith, who had been a pioneer in a number of matters of state-led economic development in the colony and who had played a central role in the recent construction of the government railway that so benefited the mines. Nonetheless, even the chief justice remained unsure about putting the mining industry at the forefront of colonial economic policy and the wider implications of the proposed bill. The scheme was at the very least controversial because of its costs, in terms of both money and manpower. After considering the provisions of the bill, on March 25, 1904, Griffith sent a lengthy response to Migeod, with various detailed points of critique. One issue cited was that the bill was imprecise in its wording, with the effect that it ultimately targeted a much larger population than only the miners, though Griffith referred to contract workers in the mines throughout his response, given that they were the apparent target of the legislation.[108] And as he saw it, the bill subjected the miners to a system that was voluntary in name but would prove to be a great barrier to their finding work in the mines in the future, because any person wishing to find a wage job in Wassa would have to acquire a certificate of registration. The certificate would be provided against a fee of one shilling, which supporters of the bill had suggested as a way of making the system self-sustaining, emphasizing that neither the state and nor the mines would shoulder any financial burden. The certificate would be handed over to an employer once a laborer had entered a contract with a company, indicating that the worker was in formal employment.[109] And it would be returned to him only after the contract had come to a legal conclusion, that is, when the employer decided that it had come to an end.[110] Griffith took grave issue with the imbalance of power that this process would entail. Expressing his concern over the extent of legal exploitative power in the hands of mine managers once they had managed to get all African contract workers in the mines to register, he concluded that mine managers were "only human" and "with a weapon in their hands as the custody of the certificate of registration it will be contrary to human nature if they and the labourer do not often disagree as to when the agreement comes to an end." He imagined a scenario wherein negligent managers and corrupt African clerks prevented men from leaving a position and finding a new one elsewhere by refusing to return their passes. He was further concerned that if the system was outsourced to African staffers, it "will constantly happen that the labourer will be kept out of his certificate."[111] Griffith predicted that only a "few managers will exercise scrupulous care in promptly restoring certificates on the termination of the agreements; but many will not trouble themselves much about it."[112]

> The matter will be often left to subordinates and there is no person more inconsiderate and more callous where a labourer is concerned that the ordinary native clerk. The laborer will ask for his certificate. He will be put off and told to come next day as the manager is away or busy or that the clerk has not got time to look through the bundle of certificates and that way 2 or 3 days will be lost.[113]

There were still too many ways for employers to take advantage of the new powers granted to them by the bill; therefore Griffith was indignant in his rebuttal: "We may eliminate one point," he declared, but "registration will not prevent labourers returning home if they wish."[114] Further, "I think we may assume that there is no objection to labourers quitting any employment at the termination of their agreement."[115] He also considered the indigenous perspective, because colonial officials were "not quite clear what view the native labourer will take of registration and how he will respond."[116] He responded to the proposal with a mixture of paternalism and fear. On the one hand, he felt a need to protect African workers, who purportedly lacked the tools to do so themselves: "No doubt provision is made in clause 23 to prevent this very thing. But the native will rarely seek the protection of this clause."[117] Griffith claimed that the African worker "with his customary experience of native ways . . . will not consider [such injustices] sufficient cause to call in the aid of clause 23."[118] He opined that the African "expects from the white man absolute justice and if the employer treats him with something less than that, and if the native has to wait about at the transport office for registration till he can be attended to, and if he meets with difficulty in getting a duplicate of his certificate, for the certificate will be often bona fide lost, he will soon throw up work or personate, and mining labour will get a bad name in the country."[119] Avoiding such an outcome would take more wide-ranging efforts than simply leaving matters in the hands of the transport office, as the bill suggested. The whole affair made Griffith lose a great deal of confidence in his fellow colonial officer Migeod, whom he now categorized as "an employer himself, who would be naturally in the side of mining companies."[120] Without some special protection for African workers besides the transport officers, and "without such a staff for carrying the scheme into execution as would guarantee its smooth working, I fear that mining labour would get this bad name."[121] It would take a specialized department to handle the passes and to protect African workers from indefinite indentureship by their employers.[122] Again, Griffith was reluctant to rely on the existing class of African clerks for this sort of task:

> I have no special reason to lay stress on this point but unless the labourers' interests are looked after by some special authority they will probable suffer in this connection and all the more because of the frequency of procedure District Commissioners will not be able to give to registration the careful attention which will be required and will have to delegate much of the work to the native clerks whose strong point is not accuracy.[123]

Any innovations to the legal framework surrounding labor had to be weighed against the risk of men avoiding mining altogether, not to mention potential negative effects on recruiting for wage work in general in the Gold Coast Colony. As Griffith explained, "It is easy to say, these fears are exaggerated but it must be admitted that registration of uneducated native labourers everywhere regarded as an experiment not to be tried without great deliberation and without ample safeguards that the remedy does not prove worse than the disease."[124]

Another point to consider was that this type of protection was not a one-off deal but would have to be guaranteed on an indefinite and ongoing basis. The chief justice provided another assessment of the bill's potential impact on colonial institutions, this time no longer adhering to the "tribal" image of the African worker. As much as Griffith complained about the obstacles in the way of traditional Africans going to court, he also admitted that "it would be highly inadvisable on the part of the government to attract the attention of the native to this clause."[125] He boasted, "At present, happily, there is hardly any litigation between employers and employed."[126] Yet he warned that the African "is by nature very litigious and it only needs encouragement to make him . . . on the lookout for causes of action against his employers."[127] Griffith foresaw that the excessive number of cases of abuse of African mining laborers eventually would force the government to expand its administration into Wassa. The chief justice advised that "before any legislation like the present bill is proceeded with the Government should give more attention to the question of whether registration of mining labour can be effectively carried out with the staff which in the existing circumstance of the colony, can be afforded."[128]

Ironically, far from wanting to eliminate African intermediaries, he suggested that more African authorities, African clerks, be employed on the mines. The Concessions Labor Ordinance targeted the desertion problem; however, in the eyes of the chief justice it employed "drastic measures" to do so.[129] Griffith had an alternative vision of a number African captains being engaged specifically to oversee African contract workers as a means of reducing desertion rates. He proposed that "a native (who would more easily recognize native faces than a European) be employed to constantly inspect labourers at the different mines, paying special attention to all newcomers."[130] Another solution to the "evil" of desertion lay in greater cooperation and less competition among the mines. "To some extent the mining companies have the remedy in their own hands; if they agree amongst themselves to lower the advance in there will be fewer peripatetic labourers."[131]

Griffith may not have been entirely unsympathetic to the free-rider problem. However, he promptly dismissed the severity of the issue for lack of evidence. Of course, this was a rather cynical tactic, considering how the lack of administration in Wassa was to blame for this problem in the first place. In the end, he contended that "the mining companies are themselves much to blame," pointing to the disaster that was the advance payment system:[132] "The transfer of services from one mine to another is due almost if not entirely to the advance systems and this transfer of services will be reduced according as advances are reduced."[133] Although he did not waste a moment considering how smaller advances would diminish general rates of recruitment; or that for many laborers sizable cash advances may have been an attraction of greatest importance; or, for that matter, that although the master and servant contract was the basis of the employer-employee relationship it was only irregularly enforced, the chief justice went to lengths to argue that the mining industry could solve its own problems: "I do not think that a case has been made out for registrations of labour, I think that for the evils so far brought to light by the miners use have remedies already to hand, that before registration of labour is seriously considered more information should be obtained."[134]

In his concluding remarks, Griffith questioned whether "anyone except the mines and the transport officer are satisfied with workability of the scheme proposed." The proposed system shared many similarities with the system of indentured Asian indentured laborers elsewhere in the British Empire; it was frank in its intention not only to punish absconders but also to hold them in the labor force in accordance with the employer's wishes. As the presiding member of the Supreme Court on the Gold Coast, Griffith found the Master and Servant Ordinance, including the amendment of 1902,[135] to be perfectly adequate. He informed Migeod that "the governor is personally opposed to the principles of the bill both in theory and practice":[136]

> I gathered that the more Sir Matthew Nathan considered the subject the less he was inclined to press the bill and I think that the Attorney General would have come to the second reading with a very open mind on the question whether the scheme would prove anyone but the mining companies and the transport officer have a good word to say to the bill.[137]

The governor's recommendation to the Colonial Secretary was that "it should not be proceeded with."[138] This opinion was further strengthened by other powerful opponents, such as the Manchester Chamber of Commerce, which as representatives of British cotton merchants in West Africa had taken to working with small-scale farmers by this time, though they had founded a joint pressure group together with the mines in the 1880s.[139]

In the end, the Concessions Labor Ordinance did not become law. Instead, mine managers had to continue trying to lower the high cost of labor with minimal colonial assistance. In 1905, the mine managers, still desiring some form of registration for African workers, agreed to build their own system from the ground up, whereby they would encourage laborers to take out certificates issued by the government transport department, and not employ anybody without one. Under the new guidelines:

1. "Any person who has engaged labour of any description and is desirous of obtaining a written contract can, at Sekondi, have the same drawn up at the transport office, where in addition every labourer will have his description recorded for the purpose of future identification."
2. "The transport office does not undertake to take any labourer who breaks his agreement but will render all the assistance we can to the employer to do so."
3. "With respect to Kroomen in particular, any person engaging a Krooman in the Colony, without ascertaining whether he had completed his agreement with his former employers, does so at his own risk, and renders himself liable to prosecution under the master and servant ordinance."
4. "The fee for signing on and rendering a man's description is 6d a head. Similar arrangements will be made at an early date for the registration of native labourers at Accra, Cape Coast Castle, Axim and Tarkwa."[140]

In view of the past suggestions of the Mines Managers' Association, it is hardly surprising that interest in the matter quickly declined so that the transport department "never issued a single form."[141] This was likely due to the system punishing the gold mines for their competitive recruitment practices rather than disciplining labor. After 1905 employers in the mines used the recruitment services of the transport department only in times of dire need. In 1908, Migeod reported, "Application is seldom made except at a time of local stringency."[142] With this turn in events the importance of the government transport department in assisting the mining industry quickly faded.

5

Labor Agents, Chiefs and Officials, 1905–1909

The Incorporation of the Northern Territories' Labor Reserve

The influx of capital into the Wassa mines reached a new low in 1905, leading many mining companies to cease production, and leaving just a few more-established and technically better-equipped ventures to prove that the industry could indeed pay its way. This period marked the end of the second gold boom in Wassa. The reduction in the number of competing mines did not, however, improve recruitment in the long term. The surplus of labor that had resulted from the closing of the smaller companies was quickly neutralized, because laborers continued to opt for work in the agricultural sector, especially cocoa farming. Therefore, within just a few months' time, the gold mines found themselves on the verge of another severe labor shortage. For 1906, Frank Cogill, the secretary for mines, reported that as anticipated, "there has been a shortage of labour, which, although not sufficient to stop the mines from working, had the effect of forcing some companies to pay higher wages than those adopted by the Mine Managers' Association."[1] This was the moment when the mines and the administration came together to collaborate on a new labor experiment, one made possible by colonial territorial expansion into the north as well as the expansion of the government railway. The discovery of labor-dense pockets in the savanna region of the north brought employers in the gold mines significant relief. Recruiting from these areas, generally directly through chiefs and colonial officials, forcibly detached wages from the demands of the market. As political promises infiltrated the recruitment machinery from the Northern Territories and became a huge motivating factor for local authorities, wage rates no longer had to adhere to the laws of supply and demand. However, in spite of all the cost-cutting measures and deception involved, this pattern of recruitment became the source of sharp political and social conflict.

This study has thus far aimed to show indirect recruitment through African labor agents as a system of voluntary bondage that could be precarious for both recruits, as well as recruiters, albeit in different measure and with different implications. It has done so in order to expand the discussion of the mobilization of African workers after the legal end of slavery, which has been dominated by the theme of colonial forced labor and other forms of mobilization similar to slavery.[2] However, other labor agents—chiefs and village headmen to be precise— were also integral to the (forced) labor mobilization system of the Gold Coast. To add further distinction between this political method of indirect recruitment and others portrayed in earlier parts of this book, this chapter focuses on some of the identifiable social and economic ramifications of the forced-labor system as it was established in the Northern Territories between 1906 and 1909. As it revisits this period when the Wassa mines recruited through chiefs, it has one main task: to compare and contrast procedures of labor mobilization under the forced-labor regime with those practiced by Liberian labor agents, the career-oriented gang leaders of the government transport office, and the more independent indigenous recruiters and supervisors. By 1909, it dawned on mining entrepreneurs that recruiting through chiefs and political officials created a cheap, yet unsustainable, labor supply, resulting in political tension in the rural areas, ill health among large section of the labor force, and mass desertion. Indigenous recruitment needed to be reinstated in its more market-driven form. But what would this new system look like?

The Emergence of the Northern Territories as a Labor Reserve

The sudden rise in prominence of the Northern Territories as a major recruiting source was not a matter of missed or overlooked opportunities. Rather, it was the byproduct of political transition. Following the War of the Golden Stool (the Yaa Asantewaa War) in 1900, which forced the Asantehene into exile, a protectorate was declared over the Northern Territories in 1902, and the region was opened up to colonial expansion.[3] However, colonization was negotiated differently here than in other parts of the colony. Disadvantageous economic conditions, including weak transportation networks and a limited amount of arable land, made any public investments in local agricultural production unlikely, maybe even unwise. However, newly appointed commissioners to the area encountered certain regions in the northeast and northwest that were inhabited by large numbers of purportedly idle young men who not long before had served as local warriors during the period of political turmoil.[4] Some groups were even described as living in an "enormously thickly populated district."[5] In consequence, an administrator explained, "the opportunity has been taken to impress upon the chiefs the advantage of sending the young men to work in the mines and the colony" with the hope that they would encourage others to do the same.[6] A labor crusade beginning in 1905 rendered these young men the protectorate's highest-valued export.[7] Just one year later, the *Report on the Transport Department* declared that "when the labor supplies of the colony are exhausted, the Northern Territory is the one place in the Gold Coast where large numbers of laborers could be obtained."[8]

The chief commissioner of the protectorate, A. E. Watherson, played a leading role in negotiations with local chiefs. He designed a model speech for a circular, issued to all officers to be delivered to the chiefs, that touched on the virtue and educational value of work and encouraged young men to become the driving economic force of this region through wage work. During these recruitment campaigns, he and other administrators made extravagant promises to local authorities regarding the wealth their young men would secure and bring back to reinvest in their home regions. For instance, it was put on official record and certainly also told to chiefs that recruits "bringing back their savings will introduce much needed money into the country."[9] Ideally, a good part of this sum would go toward cattle herding, and some would be directly acquired by the chieftaincies in the form of taxes.

Even as the officials encouraged the chiefs to exploit their authority over the male inhabitants of their villages in ways that would bring them to migrate to the mines, they were also keenly aware that the chiefs had at most a superficial acquaintance with life and work on the mines. Therefore, in a follow-up to the speeches, labor representatives from various social groups were selected to tour the mechanized mines in December 1906. After the tour, the secretary for mines gave a highly optimistic account of their visit:

> With a view to finding a new source of labour, early in December last, some 30 Labour Representatives were brought down from the

Northern Territories by the Government, so that they might see for themselves the different kinds of work and the conditions of living in the mining districts.

The labour representatives were picked from races inhabiting a large portion of the Northern Territories, and were carefully chosen so as to properly represent the districts from which they came.

The visit was a decided success, the people showing an interest in all that they saw.

On their departure, they stated that, on their return to their country, they would inform people of what they had seen, and they anticipated that their countrymen would start to come down to work in the mines.[10]

Together with the labor representatives, Cogill toured different properties in and around Tarkwa, such as the port of Sekondi and other impressive sites, which it was said left them feeling inspired enough to invest their energies in this scheme. In the following years the labor representatives assisted the chiefs in mobilizing thousands of men to work on the Wassa gold fields. Under the supervision of a district commissioner, the initial contingent of laborers, comprising a variety of social groups, arrived to work in the western protectorate in 1906. The very first group of seventeen Isalla and nine Dagarti departed for the Abbontiakoon Mines where Gerhard Stockfeld, then president of the Mine Managers' Association, was in charge.[11] Most of these men were assigned to underground mining at a low wage rate.

The experiment was immediately financially lucrative but remained structurally fragile, precisely because many of these young men "had never left their villages before."[12] They often knew little about what to expect. Also, they were initially unaware of being compensated with some of the lowest wage rates ever seen in the mining sector for the same hard work that other groups of miners were currently doing for more money. Whereas in 1903 and 1904, the Abbontiakoon mines paid its underground laborers one shilling six pence in addition to three pence on a daily basis,[13] the daily earnings of a laborer from the North for the same work was one shilling at the start of the "labor crusade." The wage rates of those men assigned to government services were even more abysmal.

Colonial officials and mine managers celebrated the scheme. According to one administer stationed in the north during the year it began, "the results have already been successful—one gang very shortly after their return proceeding to the Abbontiakoon Mines, and many others were ready to follow."[14] In the coming years, mine management's requests for laborers were swiftly answered. Indeed, when during the first decade of the twentieth century, mining entrepreneurs referred to the Northern Territories as the "best-governed portion of West Africa," they were basically heralding the triumph of the new, centralized labor-recruitment system, which was born out of the spread of British influence into the social and economic affairs of the region.[15] The labor coming out of the area was a tangible product of that influence.

The Recruitment Machinery from the Northern Territories

In 1909, the British miner James Herbert Curle, who worked both on the Rand and Wassa gold fields, insisted that in West Africa, "the relation between the government and the mining companies on the whole question of native labour are now harmonious."[16] His assessment was rooted in the recent, ongoing collaboration between mine managers and the colonial state on the labor issue. Although most migrant laborers still preferred to find work on cocoa farms in the export-crop-growing areas of the Gold Coast, as of late the mines only had to compete with one other major employer—government public works—over the rest. As Curle explained, "The natives of the hinterland can be secured only by permission of the government, but this has been given."[17] The Wassa mines could now hire a huge number of laborers at once and through a European colonial officer who facilitated much of the negotiations. The administration forwarded individual requests to particular chiefs, who, as Roger Thomas has shown, "were to recruit gangs of twenty-five men, each with a headman and musician."[18] In successful instances the names of the individual men were recorded locally and they each received a metal disc with a unique identification number before traveling south, and the chief and village headman were rewarded for their efforts. However, if things went awry or if they failed to comply with the colonial officers' requests, both the chief and village headmen could face lofty fines. The responses of local authorities to the demands for laborers were mixed, compelling political officers to employ a combination of reward and punishment to gain their support.

A number of incentives were in play. Thomas has posited that local authorities hoped to gain political favor with the British by fulfilling what they perceived as orders for laborers. In effect, responding to such calls for labor was one immediate way in which the chiefs were able to demonstrate their loyalty to Britain and to contribute to the development of the Gold Coast.[19] Chiefs in the Northern Territories, however, also received monetary compensation of around five shillings for each gang they provided. Watherson reckoned that this amount was convincing enough as an inducement for them to "use their influence to persuade the young men to go down."[20] It was also meant to compensate them for losses in labor over these several months to a year. For as the chief commissioner insisted, an offer of five shillings "makes them take more interest in the recruiting of these labourers, and compensates them for the loss of their share of the crops of the absentees as naturally less local ground in places under cultivation."[21] The chiefs also considered the decision to send off young men to the mines to be a long-term investment. As mentioned earlier, they were promised that their men would return with wealth to be reinvested in the local economy. The chiefs also had an opportunity to make additional money if their men extended their contracts with the mines. However, actual financial rewards may have been limited.

As one political official in the north recognized, it "was only through the loyalty and good will of the chiefs that we got labour for so cheap, and not because it was economic."[22] Just as the chiefs were undercompensated for labor lost in their home areas, so too was the work of the individual laborers severely undervalued and rewarded largely in a symbolic manner, if acknowledged at all, because the officials involved assumed that they would eventually return to their rural homes, where they would be supported by their families once again. Forced labor may not have been slavery. Nevertheless, it was coercive and amounted to a thinly veiled artifice. On an official level, British officials presented work in the mines as an opportunity for socioeconomic, even moral, betterment. And perhaps some recruits, who did not have access to the rich soils or minerals that would make economic and political independence a real choice, accepted it as such. Much more widespread was the official tactic of pressuring local chiefs, who in turn found strategies to compel young men to participate in the recruitment machinery.

But the greatest deception of the British did not simply lie in the fact that the actual revenue resulting from the men's engagement was generally underwhelming. Nor was it that miners from the Northern Territories were often obligated to work for a twelve-month period, in contrast to the six months that had been the norm in previous years, with most being assigned to the most dangerous underground work. It was really what happened at the end of the contractual period that decisively robbed the young men of their individual liberties.

It perhaps first needs to be explained that political officials and mine managers made certain assurances to the chiefs geared toward ensuring that all recruits were back in their respective villages in time for the farming season. They made extensive efforts to uphold the authority of the chiefs over their supposed subjects during and after the contractual period, because colonial rule itself rested on a platform of chiefly rule.[23] This emphasis is, for example, observed in the chief commissioner of the Northern Territories' insistence that men from a common village work together at a particular mine; that they maintain constant communication with authorities back home; and that a colonial official accompany them down to urban areas, in spite of the strain that this put on everyday administrative affairs in the area.[24] A commissioner escorted the men to Kumase, and then from Kumase to Tarkwa by train or in open trucks, "fitted up so that a temporary roof can be made of Willesden Canvas or corrugated iron."[25] In addition, the men were housed in secluded mining villages where they were supposed to be cut off from contact with other groups.[26] Although management contended that the separation was a means of protecting supposedly naive laborers from the north from being cheated by Akan people, this behavior likely was also driven by the need to keep them in a realm of economic ignorance to the employers' benefit. Next, political officials enforced staggered wage payments to make sure the laborers actually went back to their villages, where they would assist in farming when their contract ended. Under preexisting indigenous recruitment systems, migrant laborers who agreed to certain advance payments were tied to a labor agent until the debt was paid off. However, any authority over the men, whether economic or social, generally dissipated by the time they had completed their six- to twelve-month contract, at least under ideal circumstances, leaving the members of a gang to choose freely how and where to spend their wages. In contrast, under the new system, laborers from the Northern Territories were unable to collect the entire sum of their wages once the contract was over. Instead, they received only two fifths of what they had earned. They had to collect the remaining three fifths of the money from the local district commissioner in their home region. The secretary for mines defended this payment system as crucial for getting men "back to the country":[27] "Arrangements were made back in Tarkwa that only 6d [pence] a day should be paid to them for subsistence and clothes, and that at the end of their year they should be given two fifths to spend on goods in Coomassie and three fifths should be drafted to the Commissioners of the District to be paid [to] them on their return to their homes."[28] The behavior of these officials adds an interesting aspects to the observations of colonial commentators depicting supposedly lazy African migrants who routinely "'sit-down' in their district for a month or two, until they exhausted the profits of their labour, when they again go forth to seek employment in some other part of the coast."[29] In the case of migrants from the Northern Territories in the early twentieth century, European officials had a direct hand in their repatriation.

This form of oscillatory migration was not meant solely to appease the chiefs and maintain social control. Employers also used it as a means of advertising opportunities in the mining sector to future recruits. Returnees played a critical role in promoting work opportunities in the mines. Therefore, it was important that they returned home with more than just cash in their pockets. Exotic items from the coast truly helped to show off their new wealth: cloths, clothes, shoes, and goods that were not easily acquired back home.[30] Furthermore, these young men could expect to enjoy greater honor and esteem in their villages. In 1910 the district commissioner in Tamale, H. W. Leigh, suggested that "the wealth of the returning boys was seen and allowed time to sink into the minds of prospective mine boys."[31] In Gambaga, the commissioner noticed how "mine boys sent from here returned very satisfied with clothes and money and are apparently rather envied."[32] Another administrator observed that the "success in the North Western Province is due to the fact that the first party sent down to work remained there, brought back cloths, boxes and some savings with them, and this had led to others going."[33] To employers and officials, returnees were the best argument possible for migration, because they had seen that "the white man keeps his word."[34]

However, some men among the returnees had a different narrative to share—men who would have preferred to settle in the urban south but lacked the option of doing so. The recruitment scheme made this choice less attractive, because laborers from the Northern Territories were not encouraged to bring their wives along. This was a rather contradictory phenomenon. On the one hand, migrant laborers who bought their spouses (or families) along required additional money for travel and subsistence before the work had even begun. Cogill declared that family units "of course mean that larger money advances would have to be given to the labourers, and also if they are to be brought down by train from Coomassie, the railway charge would be increased."[35] On the other hand, he was also fully aware that the presence of wives encouraged men to work for a lengthier period of time: "If possible, labourers ought to be permitted to bring down their wives, as I am convinced that if they did so they would be willing to work for longer periods on the mines, and would keep in a better state of health."[36] Cogill assured the colonial secretary that if "the labourers are able to bring down their wives and families with them I see no reason why permanent villages should not be established."[37] All that had to be done on the company's end, as he saw it, was arranging the supply of tracts of land where the families could construct homes and grow their own foods.[38] However, officials in the north who had to deal with the complaints of the chiefs, and who were witnessing declining, or at least stagnating, economic conditions, were entirely uninterested in the proposal. Family migration became a feature of mining recruitment only in the decades to follow.[39] A more comprehensive strategy to bring more families to the mines was not undertaken until the interwar period.[40]

Layers on Indigenous Governance

Carola Lentz, who has dedicated a significant portion of her research to understanding the social and political organization of migrant workers for the Northern Territories on the mining concessions to the south, has also dealt with the roles of labor agents, especially headmen, during this early stage of recruitment. Whereas it is certain that each individual gang of twenty-five men traveled to the south in the company of a village headman, she paints quite a different picture once they arrived. According to her, "all migrants from the Northern Territories first came together under a single tribal headman, mostly a Dagomba, a member of one of the precolonial kingdoms of the north."[41] She described the number of "headmen" increasing with the number of migrants to the area. Although the conclusion that they behaved as if belonging to a single kingdom (socially) rings true for those familiar with the strict social divisions of the mines (in housing, for example), in terms of the organization of labor, this conclusion would imply a tremendous shift in the preexisting system of indirect recruitment and the organization of labor. But perhaps what had come into fruition were multiple overlapping systems of social and economic organization with temporarily indigenous supervisors, involved with work and everyday goings-on, and village headmen, who were more intricately preoccupied with the ethnopolitical matters. What can be pointed to with certainty is a persistent lack of either the desire or capital or manpower necessary to shift to a direct form of labor control at the time.

The Sick, the Rebels, the Nonreturnees: The Failure of Forced Labor

The forced labor system as it was established starting in 1905 was only a short-lived option for mine management. However, not only the frustration of employers in Wassa led to its demise. As already hinted, the actions and opinions of chiefs and laborers have to be accounted for, too. Roger Thomas has posited that because the obligation of providing cheap laborers coincided with a stream of voluntary migration to the south, chiefs were left with weaker men from whom to pick.[42] Stronger, healthier men, on the other hand, could travel to the south independently and negotiate a decent salary on their own or, more likely, as part of a group. The period leading into the first decade of the twentieth century witnessed an exodus of young men leaving their villages in the north in spite of the protests of local chiefs who, according to officials reports, "complain that they are left without sufficient labour to farm the land."[43]

The unintentional tactic of selecting from the bottom of the barrel of the labor reservoir caused frequent outbreaks in disease that spread rapidly among migrant populations, affecting the home regions as well as the mining centers. As early as 1906, one official commented that "the enlistment for labour has received a severe check, due to an epidemic (cerebro-spinal meningitis) raging in that part of the country."[44] In this context it could be fruitful to contemplate whether incidents of this kind would have been far less likely had primarily economically minded labor recruiters been involved in the process: individuals motivated by careers prospects or direct profit who carefully selected the members of their gangs; who employed financial incentives to get their men to agree to contract; and who, in doing so, incorporated a particular filtering mechanism to employment relations. In contrast, quantity was the primary focus of chiefs and village headmen from the Northern Territories.

Next to the growing frequency of requests placed and the larger number of men demanded by the mines after 1906, instances of desertion by Northern Territories laborers were on the rise.[45] Desertion rates reached a new high after 1908, when officials initiated large-scale recruitment to offset the growing number of laborers who were now choosing to look for work on railway construction rather than on the mines.[46] Recruits from the northeast absconded in especially large numbers. Of the first group of 540 men from Navrongo district that year, 200 had already absconded around Tamale. Two hundred fifty-eight men eventually reached Tarkwa district and the Abbontiakoon mines, though none remained by mid-year. Future groups of laborers also deserted—hundreds at a time, in fact.

Devastated as it was by disease and disobedience, the political method of recruitment was only briefly enticing. As the recruitment system serving the mines in Tarkwa quickly unraveled, one point of growing contention between the chiefs and political officials was the issue of losses in "tribal services" due to a growing wave of voluntary independent migration.[47] By the outbreak of the World War I, many chiefs were giving the account that thousands of men were now leaving the north and going to Kumase and the Coast, "where they can [get] better pay."[48] This trend began much earlier on. Many individuals who had completed their term of service simply chose not to go back. In 1909, of 500 laborers recruited by a Captain Warden from Gambaga in May, only 27 were left on the Abbontiakoon Mine in July. An astonishing 242 had deserted prior to reaching Tarkwa. The remaining 258 worked until the June 27, a date on which 153 men left their work. The remaining laborers later wished to return home. Despite all of Cogill's coaxing, they eventually left the area. Those who were left only worked for another twenty-six days. A considerable number of deserters never returned home. Significant numbers were recorded as never even returning to their home villages to collect the remaining three fifths of their wages. Moreover, many of the young men who did return home began to reject the authority of the chiefs. Local authorities communicated to the administration that they had to deal with growing insubordination among returned miners, who anticipated spending their hard-earned wages as they saw fit and generally preferred to spend their money in the urban areas purchasing goods that could be traded once in the north. Not only did these young men oppose local political structures upon their return (if they returned at all), but many others left for fear of being recruited by the government in the first place. The colonial state was known to employ very brutal tactics of recruitment, especially during periods of conquest and "development." Illustrating this point, a particularly repressive act of colonial "enlistment" was captured in the company records of the Asante mines from 1897, wherein a manager described government raids for laborers at a local marketplace. Supposedly, even laborers who were already in the employment of foreign employers were forced to serve as government carriers.

> In the early days of [December] we found carriers coming in freely. Some workers also came in freely—volunteers at about 150 a day, but

the [government] wanting 50 . . . raided the marketplace and caught men lawfully employed by Europeans and others fresh from the interior with produce and compelling them to leave their work and their treasured loads pressed them by force into their service as carriers.[49]

The momentary appetite for colonial expansion likely explains the brutality of this particular incident. An administrator in the north complained that not only did these types of practices deplete certain areas in the countryside at critical farming times but also that the young men were inclined to leave these areas on account of it.[50]

In the face of these frustrations, local authorities felt further aggrieved because most of the financial projections made to them by political officials never manifested. The external wealth that had been promised to them never really materialized. And the men who did return home did not have large savings to show for it. Indeed, there is no evidence that African chiefs in the Northern Territories made any meaningful financial gains from labor exports. Some chiefs eventually propositioned the state to relieve them of certain labor supply duties. Occasional food shortages were among some of the hardships the encountered, as a result. According to an administrative report from 1915, local inhabitants routinely suffered from food shortages around the beginning of the rainy season in May, and demands for food were being exacerbated by local recruitment activities. As the chief commissioner of Tamale lamented, "Food is scarce in this part of the district, three small yams costing 3d, and we are demanding too much of the people in the matter of transport etc. and making them neglect the most important task of producing food."[51] A gradual rise in the price of food was another painful consequence of the scheme.[52]

Reverting to a Nonpolitical Method of Recruitment

Starting in 1909, officials moved to restructure the recruitment system.[53] And following a meeting on January 15 of 1910 between the governor, Chief Officer Migeod, Secretary of Mines Cogill, and the acting commissioner, these three men presented a scheme "on the recruiting and employment of Northern Territory labourers."[54] The new scheme was driven by the explicit desire to eliminate the factor of force from labor mobilization in future. They wrote, "We are of the opinion that, in order to obtain voluntary labour, recruiting by the agency of chiefs and other influential persons must cease and we are also of the opinion that commissioners should take no active part in the recruiting."[55] They preferred that any employers looking to hire laborers from the Northern Territories appoint a native recruiting agent.[56] Thus, the way to reach their goal was to once again encourage individual African recruiters to do the work: "On any mining company requiring this particular class of labor [contract workers] the manager should send up a suitable headmen [labor agent] to proceed to his own country and endeavor to get his own countrymen to volunteer to accompany him back."[57] This newly regulated form of indirect recruitment combined many of the best practices of older manifestations of the system. Like the Liberian agents before them, the new agents were to be allowed to recruit only with the consent of a mining firm in the form of a letter of authority. Like village headmen from the Northern Territories, they had to visit stations on the way to the mines, checking in with the district commissioners during the process and submitting their recruits to health checks during the journey to the mines.[58] Ultimately, direct labor recruitment and control was still undesirable, and perhaps impossible, at the time.[59] Thus, indigenous recruitment and supervision was revived and institutionalized after 1910, though African intermediaries would remain under the suspicious eye of the colonial administration.

Conclusion

This study provides a number of important insights into the global labor history of imperial gold mining in Wassa, as well as in a wider West African context. It has shone a spotlight on West African male and female laborers and labor agents in the mining sector to highlight their contributions to and positions in the socioeconomic and political transformations that touched their societies during the first decades of colonial rule. Capitalist intensification evoked a variety of responses by local actors. We have seen that mining concessions in Wassa relied on a wide variety of labor relations to keep up production, including contract workers, piece workers and tributary workers. This study demonstrates these laborers' active and creative engagement with casual and contract work, and the related processes. Indigenous labor agents, responsible for bringing groups of contract men to the mines, negotiated much of the politics of labor control for European employers in the mechanized gold mines. Mine managers did not fully dictate the process of limited class formation, either. Many local and migrant laborers also worked out their potential gains and losses, using their bargaining power to develop a work schedule that fit their lifestyles and priorities. The bargaining power of African mine workers in Wassa also helped them to secure regular cash advances, high wage rates, and better treatment by their employers. They employed a variety of political and economic tools to carve out the best possible future for themselves, a future that tended to involve a continued preoccupation with agricultural production. Rarely were they wage slaves, though the coercive power of the mining firms should not be understated.

Mediators, Contract Men, and Colonial Capital takes a more nuanced view of colonial commerce by illustrating the impact of political culture on economic conditions in the Gold Coast. Although the actions of foreign mining entrepreneurs often indicated a strong intention to limit the process of industrialization (especially as it related to restricting workers' tactics), the labor market was also shaped by their lack of control and limited influence over their new surroundings. In the context of imperial gold mining, mining magnates encountered different socioeconomic and political obstacles in different parts of the globe. In

Wassa, in particular, their inability to capture the power of the state created openings for African intermediaries, who assisted them by bringing laborers to the mines. And in that sense, this global labor history has also dealt with the uneven spread of globalization in Africa.

On a general level, the history of colonial mines in West Africa in the nineteenth and twentieth centuries stands out for the high number of indigenous promoters and managers involved. This study has extended the debate on African intermediaries in colonial commerce by pointing out how different types of labor agents maintained different types of relationships with the mines. Although these figures may have emerged in response to capitalist intensification, their loyalty to that project was not always straightforward.

This study provides a complex view of life and work in the urbanizing mining centers, with all of their attractions and trappings. As far as groups of contract workers were concerned, the excitement of living in Wassa and gaining equal political status with other African men there, was thrilling. However, not all mines were created equal. Some provided better entertainment or greater access to foodstuffs from home. In addition, the character of the African and European supervisors on duty could intimately shape the experiences of individual laborers. Furthermore, for contract men, the danger of underground work was compounded with the insecurities of living in ethnically segregated mining towns where they had to fend for themselves and their own, forcing them to strengthen what were initially tenuous bonds with others who shared a common language. In connection with that point, *Mediators, Contract Men, and Colonial Capital* captures the dynamic relationship that existed between African labor agents and contract men, analyzing their interactions in the context of the diverse relationships of dependence that grew out of the legal abolition of master-slave relationships on the Gold Coast and most parts of West Africa. Credit relationships spread to the modern sector. Although they came with offerings of social security similar to employment relationships in the past, they were not meant to enforce permanent or long-lasting arrangements. However, this is not to say that debt was never used coercively in and around the mining centers. Although the prospect of gaining wealth was emphasized with the initial advance payment employed to entice a mine worker, the ambition of becoming rich became progressively less tangible for those laborers digging themselves deeper and deeper into a web of debt, as they spent months waiting for the disbursement of their wages due to the extended pay system. Indebtedness with the mining firms, with the leaders of their gang, and especially with local Akan moneylenders could have a detrimental impact on their ability to approve of and improve the work conditions to which they were subjected. This study also suggests that the mine workers' labor power could be exploited by moneylenders who would sign them up with another indigenous or foreign company at the end of their initial contract so that they could pay off a debt. Gambling was the source of others' financial woes. Another threat came from bandits who lurked along frequented migration routes. Although many laborers in the Wassa mines managed to save money, return home in honor, and become living testimony of the potential of wage work, others floundered and struggled to retain their autonomy and power.

It is hoped that the research presented here will stimulate further inquiries related to socioeconomic and political change in Wassa during the late nineteenth and early twentieth centuries. Out of its findings more research can be launched, examining the agency of African workers and shifts in their bargaining power during the colonial period; credit, compulsory labor, and voluntary bondage during the colonial period; and workers' consciousness outside of working-class formation. Last, the changing professional and social roles of women mine workers in colonial Africa demand growing research.

Autonomy, Mobility and the Fluidity of the Early Colonial West African Labor Market

Other studies of the early colonial Wassa gold mines have shed light on the diversity of the labor market in Wassa in terms of its demographics and labor relations. However, the logistics behind the creation of that labor market for the mining sector have not been elaborated on. The movement of laborers between West African regions or colonies, often with the help of labor agents, required financial and political wrangling and negotiation. It entailed, among other things, transportation along sea and overland routes. Thus, research on the European vessels traveling along the West African coast could reveal a great deal of information not just about the laborers in their employ but also about those moving across great distances for work, illustrating one particular facet of labor migration globally.

Colonial records were unable to capture in any precise manner the movement of people between various states in West Africa during this period, whether overland or by sea. Yet we know that opportunities in and around Wassa brought together laborers and labor agents from the Gold Coast, Liberia, Nigeria, and Sierra Leone, to name a few places.

Traders also swarmed the area to sell goods. The factor of having migrated often weakened a laborer's bargaining power in a concrete way. A special kind of precarity associated with migration stemmed from the money that had already been spent to get them to Wassa in the first place. The money that was required to get them home, the legal barriers of the Master and Servant Ordinance, and cultural isolation in the mining centers were other hindrances. It is hardly a coincidence that local labor preferred and managed to work in a casual manner, although certain cultural impediments added to a general dislike for the underground work to which contract workers commonly agreed. A great deal of additional information will come from studies that take these movements and the social networks they fostered in colonial workplaces, as well as in agriculture, into account.

In many parts of empire, control over labor shaped the manner and degree of a region's integration into the international market. The Wassa mines benefited from a fluid and diverse labor market during the late nineteenth and early twentieth centuries, when border controls were weak or nonexistent in many part of West Africa, creating openings for individuals in desperate need of better economic status. Yet the role of state actions, border and regulations has not been underestimated, either. Starting in the 1890s, Wassa was also deeply affected by the economic policies of France and Liberia. The consequences of territorial encroachment and the loss of access to Liberian contract men were far reaching, as this study has shown. Thinking about the limits of globalization is essential to understanding African economic history in regional and other terms.

Indigenous Responses to Economic Changes

The shortage of men working under contract and the habitual movement of men in and out of the wage-labor market that characterized both the mining sector and the larger wage labor market in West Africa during the early colonial period created openings for labor agents with a variety of motivations and financial means. They were novel figures, but the mechanisms that they used were adopted from the past. Security between the recruiter and recruits was, however, rooted in short-term credit relationships rather than in any form of ownership or coercion. Different types of labor agents seem to have dominated different historical periods. Liberian gang leaders, who were prevalent during the nineteenth century, relied on letters of authorization and capital from their mining firms to bring gangs from Liberia or Sierra Leone to the Gold Coast. Their relationship was hard earned and proven with the help of the "good books" they received from a succession of European employers. Their future employment prospects generally also rested with these employers, given a lack of economic opportunities and combative relationship with leaders in their home regions. The career-mindedness and maneuvering of this group deserves attention in its own right, for they were not simply lackeys of foreign capital.

By the twentieth century, the changing composition of the population of contract workers in Wassa reflected the novel relationships that managers in the mechanized mines in Wassa were forced to forge with other men offering their services as labor agents. Labor agents from other parts of West African had access to different ethnic groups in large numbers, further diversifying the labor market for the mines. Moreover, recruiters from the southern Gold Coast had easy access to men who had completed their contracts in this and other economic sectors near the port of Sekondi. Alternatively, the new labor agents managed to lure some of them into new positions before the completion of a contract. Also, in an illustration of the darker sides of the spread of credit relationships throughout West Africa, Akan moneylenders were able to instrumentalize the debts of mineworkers to get them to agree to gang labor.

The professional profiles of the Hausa labor agent Madam Mariam and, to some extent, of the European agent Eamonson, demonstrate that labor agents were not just sketchy moneymakers but that they could also have long-term plans and success. They show that there was wealth to be made in this activity and that the promise of it could draw in actors from faraway places. This study reveals that these regional activities made possible larger transformations, for instance, those connected to colonial conquest and the spread of capitalism. Therefore, developing a fuller picture of African entrepreneurial business activities during this stage helps to provide a more complete understanding of the collaborations between Africans and European that paved the way for such processes. Resistance to capital does not tell the full story. Nor does the notion that Africans were generally coerced into wage work.

Politics, Workers Tactics, and Socioeconomic Change

Through a number of studies produced in the field of global labor history, labor intermediaries in industries of scale are now recognized as a phenomenon of the global labor force, which emerged with empire and capitalist intensification worldwide.[1] Scholars have debated the origin of these figures, their roles and practices, and the reason for the emergence and persistence of indirect recruitment, especially in many parts of the global South in particular. Yet although research on these figures encourages a great deal of south-south historical exchange and comparison, admittedly there remains a need to engage previous research that fully embraced the Eurocentric model of free wage labor with all of its ideals.

This study uses court records as a means of understanding the concerns and politics of laborers at the lowest level of the mining hierarchy. Economic transformation left the bulk of the politics of labor to continue to play out among Africans. It is therefore not surprising that the majority of case records on the Master and Servant Act for the period under investigation highlight disputes between gang leaders and laborers. Whether they concerned pay, punishment, or the provision of services, the actors involved generally were African. The mine managers, occasionally to their advantage and at other times to their disadvantage, had created a metaphorical moat that held laborers' concern at bay.

As a colonial agent sending African laborers to work in the mines, the British chief of the transport office, F. W. H. Migeod also created new avenues through which laborers could defend their rights against abusive managers and supervisors. However, these correspondences between the agency and its employees also seem to have come to fruition only in the most desperate cases, particularly when laborers had a loss of income at stake. Otherwise, it was probably easiest and most gratifying for them to leave work without notice and hope not to get caught, given the paucity of structures put into place to find deserters along the coast and the irregular implementation of labor laws by the state.

Outside of the British courts and agencies, and in the absence of widespread class formation, workers still found ways to exercise their power in front of employers. This study has shown that despite not having individual contracts, casual laborers (that is female casual laborers) engaged in collective action for adequate pay. The politics of particular gangs of laborers could easily spread to other workers.

Women's Work and the Mechanized Mines

Women have not taken up a large space in the history of colonial mining in Africa. Nonetheless, this study has tried to make up some ground in that arena. Women workers had particular skills that they could monetize, and managers in the Wassa mines seems to have noticed that quickly. But women were not simply relegated to applying the same skills, panning gold in particular, in both the small-scale and large-scale mining sectors. Rather, their tasks, which developed out of precolonial gender divisions in mining, also transformed with the introduction of more advanced machinery after the 1890s. Using the colonial African context as an example, wherein technological advancement could sometimes demand even greater labor-intensive investments, it is perhaps more accurate to think about how skills changed along with advanced technology historically (and were not necessarily made obsolete by it).

The current study has shed a new light on the colonial Wassa mines and the diverse group of male and female African laborers, labor agents and creditors, and the political tug-of-war that made possible an expansion of the labor market. It pays deepest respect to a dynamic group of individuals who in most cases receive limited, derogatory, or no mention in contemporary writings. Their histories, and the histories of workers all over Africa deserve further attention with an eye for detail, illuminating the economic and social negotiations they entailed.

Notes

Introduction

1. Construction on the Sekondi–Tarkwa line started in 1898.
2. Mary Gaunt, *Alone in West Africa* (New York: Charles Scribner's Sons, 1911), 121–22.
3. Gaunt, *Alone in West Africa*, 122.
4. Raymond E. Dumett, "Gold: Akan Goldfields: 1400 to 1800," in *Encyclopedia of African History*, ed. Kevin Shillington (New York: Taylor & Francis Group, 2004), 1:586–87.
5. Raymond E. Dumett, "Parallel Mining Frontiers in the Gold Coast and Asante in the Late 19th and Early 20th Centuries," in *Mining Frontiers in Africa*, ed. Katja Werthmann and Tilo Grätz (Köln: Rüdiger Köppe Verlag, 2012), 35.
6. See Dumett, "Gold Mining and the State in the Akan Region in the Pre-Colonial Period," in *Research in Economic Anthropology: An Annual Compilation of Research*, ed. George Dalton (Greenwich, CT: JAI Press, 1979), 2:37–68.
7. Dumett, *El Dorado in West Africa: The Gold-Mining Frontier, African Labor, and Colonial Capitalism in the Gold Coast, 1875–1900* (Athens: Ohio University Press, 1998), 264–69 and 284–87.
8. For a compilation of scholarship in African history with a central theme of class formation, see Frederick Cooper, "African Labor History," in *Global Labor History: A State of the Art*, ed. Jan Lucassen (Bern: Peter Lang AG, 2006), 91–116.
9. However, in the debate over the "invention of tradition," some scholars have argued that even innovative means of social organization during the colonial period were related to something very real, something familiar to African people and somehow—though often not intensely—connected to their past.
10. See, for example, Donald Denoon, *South African Mining* (unknown binding, 1982); Stanley Trapido, "South Africa in a Comparative Study of Industrialization," *Journal of Development Studies* 7, no. 3 (1971): 311.
11. With a focus on Portuguese African colonies, Alexander Keese has suggested that, starting in the interwar period and leading into the era of decolonization, the colonial state experimented with an array of ideas, including taxation, vagrancy laws, and compulsion, to increase and manage

the local labor force for agricultural production and infrastructural development. Keese, "Slow Abolition within the Colonial Mind: British and French Debates about 'Vagrancy,' 'African Laziness,' and Forced Labour in West Central and South Central Africa, 1945–1965," *International Review of Social History* 59, no. 3 (2014): 377–407; Keese, "Forced Labour in the 'Gorgulho Years': Understanding Reform and Repression in Rural São Tomé e Príncipe, 1945–1953," *Itinerario* 38, no. 1 (2014): 103–24. Indeed, labor histories stretching beyond the boundaries of southern Africa have illuminated cases of state-sponsored compulsory labor all over the continent, proposing its widespread and systematic relevance in regions such as Equatorial Guinea, eastern Nigeria, and Senegal, to name a few locations. Ibrahim Sundiata, *From Slaving to Neoslavery: The Bight of Biafra and Fernando Po in the Era of Abolition, 1827–1930* (Madison: University of Wisconsin Press, 1996); Babacar Fall, *Social History in French West Africa: Forced Labour, Labour Market, Women and Politics* (Amsterdam: Sephis, 2002); Enrique Martino, "Clandestine Recruitment Networks in the Bight of Biafra: Fernando Pó's Answer to the Labour Question, 1926–1945," *International Review of Social History* 57 (2012): 39–72. Carolyn A. Brown, *"We Were All Slaves": African Miners, Culture, and Resistance at the Enugu Government Colliery* (Cape Town: James Currey, 2003). Dumett, *El Dorado in West Africa*, 16.

12. Roger Thomas, "Forced Labor in British West Africa: The Case of the Northern Territories of the Gold Coast 1906–1927," *Journal of African History* 14 (1973): 79–103.

13. Anne Phillips, *The Enigma of Colonialism: British Policy in West Africa* (London: James Currey, 1989).

14. Kenneth Swindell and Alieu Jeng, *Migrants, Credit and Climate: The Gambian Groundnut Trade, 1834–1934* (Leiden: Brill, 2006), 120. In response, Gareth Austin has suggested that the experience of abolition in the Caribbean, where former slaves preferred to work for themselves, only strengthened officials' doubts as to whether free labor was necessarily more productive. Austin, "Cash Crops and Freedom: Export Agriculture and the Decline of Slavery in Colonial West Africa," *International Review of Social History* 54 (2009): 1–37.

15. For private-sector recruitment through chiefs in colonial West Africa, see Thomas, "Forced Labour in British West Africa"; Bill Freund, *Capital and Labour in the Nigerian Tin Mines* (Atlantic Highlands, NJ: Longman, 1981); Brown, *"We Were All Slaves"*; Carola Lentz, *Ethnicity and the Making of History in Northern Ghana* (Edinburgh: Edinburgh University Press, 2006), 138–43.

16. For studies on the role of traditional African authorities, in particular chiefs, in labor mobilization for various government services in colonial Ghana, see Kwabena O. Akurang-Parry, "Colonial Forced Labor Policies for Road-Building in Southern Ghana and International Anti-forced Labor

Pressures, 1900–1940," *African Economic History* 28 (2000): 1–25; Akurang-Parry, "'The Loads Are Heavier Than Usual': Forced Labor by Women and Children in the Central Province, Gold Coast (Colonial Ghana), c. 1900–1940,'" *African Economic History* 30 (2002): 31–51; Akurang-Parry, "African Agency and Cultural Initiatives in the British Imperial Military and Labor Recruitment Drives in the Gold Coast (Colonial Ghana) during the First World War," *African Identities*, 4 (2006): 213–34. See also Peter Geschiere "Chiefs and Colonial Rule in Cameroon: Inventing Chieftaincy, French and British Style," *Africa: Journal of the International African Institute* 63 (1993): 151–75; and David Killingray "Labour Exploitation for Military Campaigns in British Colonial Africa 1870–1945," *Journal of Contemporary History* 24 (Jul., 1989): 483–501; Killingray, "Labour Mobilization in British Colonial Africa for the War Effort, 1936–46," in *Africa and the Second World War*, ed. David Killingray and Richard Rathbone (London: Palgrave Macmillan, 1986), 68–96.

17. Some chiefs have been shown to have responded to the colonial state's demands for labor to increase their individual power, but in other cases not complying could result in fines.

18. In Wassa, the reliance on indigenous authorities shaped the manner in which land rights were negotiated during the first mining boom: "Even though the European gold seekers and diggers in West Africa had the same aggressive propensities as their counterparts in North America or Australia, and sought to take advantage of kings and chiefs over the grant of mineral concessions at every turn, they could not sweep away African land rights, and they soon became aware that colonial authorities on the coast took at least a nominal responsibility that expatriate mining leases were bona fide legal instruments with provisions for consideration and the payment of regular rights to chiefs." Dumett, *El Dorado in West Africa*, 16.

19. Thomas, "Forced Labour in British West Africa."

20. Disregarded in African economic history for quite some time, the theory of the "backward bending supply curve" was revisited in 2014, leading to the conclusion while the system of ideas behind the theory might have been applicable to certain labor dense pockets of colonial West Africa, it certainly did not explain the dynamics of the wage labor market within the broader region. Austin, "Vent for Surplus or Productivity Breakthrough? The Take-off of Ghanaian Cocoa Exports, c. 1890–1936," *Economic History Review* 64 (2014): 1–30. For an original voice on the theory, see Max Weber, *General Economic History*, trans. Frank H. Knight (New York: Greenberg Publishers, 1927; repr., New York: Dover, 2012.

21. "Only a Kru-boy, but . . . ," *Work and Workers in the Mission Field* (Wesleyan Methodist Missionary Society, 1920), p. 23.

22. In particular, Frankema and van Waijenburg have measured steady growth in real wages between the 1910s and the 1960s. Ewout Frankema

and Marlous van Waijenburg, "Structural Impediments to African Growth? New Evidence from Real Wages in British Africa, 1880–1965," *Journal of Economic History*, 72, no. 4 (December 2012): 895–926.

23. Frankema and van Waijenburg, "Structural Impediments."

24. See Jeff Crisp, *The Story of an African Working Class: Ghanaian Miners' Struggles, 1870–1980* (London: Zed Books, 1984); Crisp, "Productivity and Protest: Scientific Management in the Ghanaian Gold Mines, 1947–1956," in *Struggle for the City: Migrant Labor, Capital, and the State in Urban Africa*, ed. Frederick Cooper (Beverly Hills, CA: SAGE, 1983), 91–130.

25. Peter Alexander, "Challenging Cheap-Labour Theory: Natal and Transvaal Coal Miners, ca. 1890–1950," *Labor History* 49 (2008): 47–70. Dumett has shown that the narrative of cheap-labor theory does not fit the case of nineteenth-century colonial Ghana. He has convincingly argued that there is no indication that mine managers deliberately constructed a nonpermanent migrant labor force to lower wages and to enhance labor control: "Rather they relied on it because it was the only alternative." Dumett, *El Dorado in West Africa*, 224–25, 233, 269–70.

26. Dumett, "Parallel Mining Frontiers in the Gold Coast and Asante," 35.

27. Dumett, "Parallel Mining Frontiers," 38.

28. Dumett, *El Dorado in West Africa*, 144–52.

29. Jane Burbank and Frederick Cooper, "Imperial Trajectories," in *Empires in World History: Power and the Politics of Difference* (Princeton, NJ: Princeton University Press, 2010), 14.

30. Benjamin N. Lawrance, Emily Lynn Osborn, and Robert L. Richard, "Introduction: African Intermediaries and the 'Bargain' of Collaboration," in *Intermediaries, Interpreters, and Clerks: African Employees in the Making of Colonial Africa* (Madison: University of Wisconsin Press, 2006), 4.

31. See Marcel van der Linden, "The Origins, Spread and Normalization of Free Wage Labour," in *Free and Unfree Labour: The Debate Continues*, ed. Tom Brass and Marcel van der Linden (Bern: Peter Lang AG, 1997), 501–23.

32. There is a long, if not prominent, tradition of scholars emphasizing the role of unfree labor in capitalism predating the emergence of global labor history, leading all the way back to Marx himself.

33. Gervase Clarence-Smith, "Thou Shalt Not Articulate Modes of Production," *Canadian Journal of African Studies / Revue Canadienne des Études Africaines* 19 (1985): 19–22.

34. Dumett, *El Dorado in West Africa*, 18–19.

35. A number of scholars have begun to theorize about the place of Africa in global labor history: Carolyn A. Brown and Marcel van der Linden, "Shifting Boundaries between Free and Unfree Labor: Introduction," *International Labor and Working-Class History* 78, no. 1 (2010): 4–11;

Franco Barchiesi and Stefano Bellucci, "Introduction," *International Labor and Working-Class History* 86 (2014): 77–84; Karin Hofmeester, Jan Lucassen, and Filipa Ribeiro da Silva, "No Global Labor History without Africa: Reciprocal Comparison and Beyond," *History in Africa* 41 (2014): 249–76. For literature by scholars of Africa emphasizing the limits of globalization, see Cooper, *Colonialism in Question: Theory, Knowledge, History* (Berkeley: University of California Press, 2005), 91–112; James Ferguson, *Global Shadows: Africa in the Neoliberal World Order* (Durham, NC: Duke University Press, 2006); Franco Barchiesi, "How Far from Africa's Shore? A Response to Marcel van der Linden's Map for Global Labor History," *International Labor and Working-Class History* 82, no. 3 (2012): 77–84.

36. Due to the work being done at institutions such as the International Institute for Social History in Amsterdam and the IGK Work and Human Lifecycle in Global History in Berlin, the number of scholars actively researching and illuminating the compatibility of unfree and nondispossessed wage labor within the capitalist mode of production has increased significantly since the 1990s. Relevant readings include the following: Tom Brass and Marcel van der Linden, eds., *Free and Unfree Labor: The Debate Continues* (Bern: Peter Lang AG, 1997); Andreas Eckert, "What Is Global Labour History Good For?" in *Work in a Modern Society: The German Historical Experience in Comparative Perspective*, ed. J. Kocka (New York: Berghahn Books, 2013), 169–82; van der Linden, "Plädoyer für eine Historische Neubestimmung der Welt-Arbeiterklasse," *Sozial Geschichte: Zeitschrift für historische Analyse des 20. und 21. Jahrhunderts* 20 (2005): 7–28; van der Linden, "Warum Gab (und Gibt) es Sklaverei im Kapitalismus? Eine einfache und dennoch schwer zu beantwortende Frage," in *Unfreie Arbeit: Ökonomische und Kulturgeschichtliche Perspektiven*, ed. M. Erdem Kabadayi and Tobias Reichardt (Zürich: Georg Olms, 2007): 260–79.

37. Hill describes the granting of credit by members of the community (and not professional moneylenders) as both a profit-making venture and a civic duty in certain villages of rural West Africa before and during the colonial period. A similar form of agrarian credit was also a widespread phenomenon in England before the Industrial Revolution. Polly Hill, *Development Economics on Trial: The Anthropological Case for a Prosecution* (Cambridge: Cambridge University Press, 1986), 83–84.

Chapter One

1. See, for example, Paul Rosenblum, "Gold Mining in Ghana 1874–1900" (PhD diss., Columbia University, 1972).

2. The construction of a government railway only made it to the top of the mining companies' list of demands around 1883.

3. Dumett, *El Dorado in West Africa*, 11.

4. Richard F. Burton and Verney L. Cameron, *To the Gold Coast for Gold: A Personal Narrative* (London, 1883), 2:298.

5. Marie J. Bonnat, *Marie-Joseph Bonnat et les Ashanti, 1869–1874* (Paris: Société des Africanistes, 1994).

6. Friedrich August Ramseyer and Johannes Kühne, *Four Years in Ashantee* (Summit, NJ: J. Nisbet & Co, 1878), 58–59.

7. E. T. McCarthy, "Early Days on the Gold Coast," *Mining Magazine* 1 (December 1909): 291–94 (quote from 292).

8. McCarthy, "Early Days," 292.

9. Dumett, *El Dorado in West Africa*, 213, table 7.1.

10. Dumett, *El Dorado in West Africa*.

11. Ramseyer and Kühne, *Four Years in Ashantee*, 58–59.

12. Dumett, *El Dorado in West Africa*, 11.

13. Dumett, *El Dorado in West Africa*, 109–16.

14. Dumett, *El Dorado in West Africa*, 111.

15. McCarthy, "Early Days on the Gold Coast," 72.

16. McCarthy, "Early Days on the Gold Coast," 80.

17. Table 4.1 in Dumett, *El Dorado in West Africa*, 107–8 gives an overview of the major operating mining companies with concessions between 1887 and 1895.

18. Richard F. Burton, letter to the editor, *Mining World and Engineering Record*, March 17, 1883, p. 283.

19. Burton, letter to *Mining World*.

20. Burton and Cameron, *To the Gold Coast for Gold*, vol. 1.

21. Dumett, *El Dorado in West Africa*, 24, 164.

22. Ferdinand Fitzgerald, "The Gold Coast Colony Roads and Seaport," *African Times*, June 1, 1875, 69.

23. James Africanus Beale Horton, "The West Africa Gold Fields," *African Times*, December 1, 1877, 139.

24. Dumett, *El Dorado in West Africa*, 100–1.

25. In spite of the expansion of the Asante kingdom in the eighteenth century, the Fante people of the coastal region of colonial Ghana were in a unique position to shape the terms of British expansion on the Gold Coast, which in turn put them in a strong political and military position during British colonial rule. Nevertheless, self-governance was their ulterior motive when it came to cooperating with the British. Founded in 1868, the Fante Confederation, which was run by prominent Fante merchants and politicians, asserted the will to rule the Fante people (including all Twi-speakers, their spoken languages being mutually intelligible) by means of national Fante law. Rebecca Shumway, *Fante and the Transatlantic Slave Trade* (Rochester, NY: University of Rochester Press, 2011).

26. Mr. Fitzgerald to the Earl of Kimberley, March 2, 1872, CO 879/4, no. 49, British National Archives (hereafter PRO).

27. Prior to 1883, the colonial government was pressured by mining companies to develop a waterway leading from the coast to where the Ankobra and Bonsa rivers join. From there, a short land road would lead to Tarkwa.

28. Fitzgerald, "The Gold Coast Colony Roads and Seaport," 68.

29. Fitzgerald, "The Gold Coast Colony Lands," *African Times*, April 1, 1875, 43.

30. Fitzgerald, "The Gold Coast Colony Lands," 43.

31. Fitzgerald, "The Gold Coast Colony Lands," 43.

32. Fitzgerald, "The Gold Coast Colony Lands," 43.

33. Fitzgerald, "The Gold Coast Colony Lands," 43.

34. Horton, "The West Africa Gold Fields," 139.

35. Horton, "The West Africa Gold Fields," 139.

36. Horton, "The West Africa Gold Fields," 139.

37. Horton, "The West Africa Gold Fields," 139.

38. Dumett, *El Dorado in West Africa*, 166–68.

39. Second Monthly Report on the Tarquah District, May 31 1882, Enclosure in No. 8 in [C 3687], British Parliamentary Papers (hereafter PP) 1883, 20.

40. Kwamina B. Dickson, *A Historical Geography of Ghana* (Cambridge: Cambridge University Press, 1969), 251.

41. Public Records and Archives Administration Department of Ghana (hereafter PRAAD), ADM 12/3/1, no. 26, 94.

42. Henry Higgins was temporarily appointed district commissioner of Tarkwa during Commissioner Cuscaden's eight months of sick leave.

43. PRAAD ADM 12/3/1, 95.

44. PRAAD ADM 12/3/1, 95.

45. PRAAD ADM 12/3/1, 95.

46. PRAAD ADM 12/3/1, 95.

47. Dumett, *El Dorado in West Africa*, 105, 234–38.

48. PRAAD ADM 12/3/1, 96. After 1900, it happened on several occasions that British employees, who had been dismissed from jobs with a gold mining company in Wassa, were left destitute and had to ask the colonial state to cover the cost of their passages to England. The deportation of these "undesirables" placed a heavy burden on state expenses leading into the second boom period. See Gold Coast Departmental Report for the Year 1903, PRAAD ADM 5/1/12; Gold Coast Departmental Report for the Year 1904, PRAAD ADM 5/1/13; Decima M. Guggisberg, *We Two in West Africa* (New York: W. Heinemann, 1909), 124–25.

49. Report by Civil Commissioner Cuscaden, August 31, 1881, PP 1883, Enclosure 1 in No. 1 in [C 3687], 1.

50. Report by Civil Commissioner Cuscaden, January 31, 1882, PP 1883, Enclosure 3 in No. 1 in [C 3687], 3.

51. Report by Civil Commissioner Cuscaden, PP 1883, Enclosure 3 in No. 1 in [C 3687], 3.

52. Report by Civil Commissioner Cuscaden, December 31, 1881, PP 1883, Enclosure 2 in No. 1 in [C 3687], 3; PP 1883, Enclosure in No. 8 in [C 3687], 20.

53. Dumett, *El Dorado in West Africa*, 103–4.

54. "Effuenta Gold Mines Company to District Commissioner Tarkwa District." PRAAD ADM 11/1/845.

55. PP 1883, Enclosure 2 in No. 1 in [C 3687], 3.

56. PP 1883, Enclosure 2 in No. 1 in [C 3687], 3.

57. PP 1883, Enclosure 2 in No. 1 in [C 3687], 3.

58. PP 1883, Enclosure 3 in No. 1 in [C 3687], 4.

59. PP 1883, Enclosure 3 in No. 1 in [C 3687], 4.

60. PP 1883, Enclosure 3 in No. 1 in [C 3687], 4.

61. Dumett, *El Dorado in West Africa*, ch. 3.

62. Dumett, *El Dorado in West Africa*, 168.

63. Dumett, *El Dorado in West Africa*, 164.

64. Dumett, "Obstacles to Government-Assisted Agricultural Development in West Africa: Cotton-Growing Experimentation in Ghana in the Early Twentieth Century," *Agricultural History Review* 23 (1975): 159.

65. PP 1889, No. 66 in [C 5620–24], 14.

66. District Commissioner's Tours in Outlying Portions of Tarquah District, PRAAD ADM 11/1/845, No. 31/1923.

67. Robert Szereszewski, *Structural Changes in the Economy of Ghana, 1891–1911* (London: Weidenfeld & Nicolson, 1965), 38.

68. Guggisberg, *We Two in West Africa*, 1:99.

69. See my forthcoming article "Dreams of a 'Johannesburg of West Africa': The Gold Coast's Moment in the Imperial Rush for Gold" for an elaboration of promotional activities around West African mining at the turn of the twentieth century.

70. Crisp, *Story of an African Working Class*, 21.

71. G. Keith Allen, "Gold Mining in Ghana," *African Affairs* 57 (1958): 224.

72. "The Gold Coast, West Africa," *Kalgoorlie Miner*, September 17, 1901, 3.

73. Gold Coast Departmental Report for the Year 1902, PRAAD ADM 5/1/11.

74. Dredging on the Ankobra, Fura, and Offin Rivers also accelerated during this period. It reached a peak in 1909, when fifteen dredges were operating and 20,102 ounces of gold recovered. PRAAD ADM 5/1/18, Gold Coast Departmental Report for the Year 1909.

75. Decima Moore Guggisberg, *We Two in West Africa* (New York: W. Heinemann, 1909), 1:127.

76. "West African Results," *Economist*, January 3, 1903, 14.

77. Census of the British Empire for 1901, PP 1905, [Cd. 2660], 145.

78. MS 7ss.12 s.1, 4, Rhodes House Library (hereafter RHO), Oxford.

79. Dumett, *El Dorado in West Africa*, 122.

80. Guggisberg did not mince words when describing a miner of the old class, a remnant from the first gold boom, as a "ragged, unshaven and unwashed, hard put to it. His company having failed, and the mine where he was employed having closed down, he was left tramping the 'bush' in search of work." *We Two in West Africa*, 1:143.

81. Guggisberg, *We Two in West Africa*, 1:143.

82. Guggisberg, *We Two in West Africa*, 1:127.

83. David Dorward's analysis of two brothers who worked as miners in Victoria, Australia, and later migrated to the Wassa gold fields highlights the racial tensions and inequalities pervasive in large-scale West African mining sector in the early twentieth century: "'Nigger Driver Brothers': Australian Colonial Racism in the Early Gold Coast Mining Industry," *Ghana Studies*, 5 (2002): 197–214. In contrast, in his observations of the wider colony, Dumett has contended that racial tensions hardly increased during the nineteenth and early twentieth centuries. *El Dorado in West Africa*, 355–45.

84. Guggisberg, *We Two in West Africa*, 1:99–100.

85. "Death of Mr. Percy Tarbutt."

86. "Death of Mr. Percy Tarbutt," *Financial Times*, June 1, 1904, 5.

87. "In the Matter of the Companies Act 1862 and 1867 and the Matter of Tarbutt's Liquid Fuel Company Limited," *London Gazette*, June 7, 1887, 3094.

88. Ian Phimister and Jeremy Mouat, "Mining, Engineers and Risk Management: British Overseas Investment, 1894–1914," *South African Historical Journal* 49 (2003): 3, 15.

89. "West African Finance," *Economist*, December 24, 1904, 2095.

90. "West African Finance."

91. Walter H. Wills and R. J. Barrett, *The Anglo-African Who's Who and Biographical Sketchbook* (London: George Routledge & Sons, Ltd., 1905), 195.

92. Phimister and Mouat show that shareholder market manipulation continued to be an issue after 1900, and that Tarbutt was an especially ruthless actor. "Mining, Engineers and Risk Management," 4.

93. See also "Fanti Consolidated Mines, Limited," *Economist*, February 16, 1901, 249.

94. Percy C. Tarbutt, [title unknown], *African Review of Mining, Finance and Commerce*, March 2, 1901. Cited in Phimister and Mouat, "Mining, Engineers and Risk Management," 13.

174 *Notes to pp. 43–47*

95. Tarbutt, [title unknown].

96. "Abbontiakoon (Wassaw) Mines Limited," *Economist*, May 4, 1901, 675.

97. "Abbontiakoon (Wassaw) Mines Limited." In 1903, part of the property was turned into a subsidiary called the Abbontiakoon Block 1. However, the subsidiary company gradually subsumed the land of the Abbontiakoon (Wassaw) Company and in 1909 fully absorbed it.

98. The brothers Louis and David Gowan, innovative and skilled engineers, were pioneers in the introduction of more modern, capital-intensive, crushing machinery. PP 1883, enclosure 1 in no. 19 in [C 3687], 54.

99. Guggisberg, *We Two in West Africa*, 1:130–31.

100. Guggisberg, *We Two in West Africa*, 1:130.

101. "Gold Coast Investment Company Limited," *Economist*, March 23, 1901, 452.

102. "Abbontiakoon (Wassaw) Mines Limited," 675.

103. Guggisberg, *We Two in West Africa*, 1:135.

104. Guggisberg, *We Two in West Africa*, 1:138–39.

105. This process was replaced with amalgamation around 1909, which gave way to "a mill of 25 stamps and 2 tube-mills . . . together with sand and slime plant," in 1912. Guggisberg, *We Two in West Africa*, 138–39.

106. Report on the Mining Industry for the Year 1905, s. 1, PRO CO 98/14.

107. Report on the Mining Industry for the Year 1905, s. 5.

108. Manager F. Stone to the Secretary for Mines, June 15, 1903, London Metropolitan Archives (hereafter LMA) MS14171, vol. 9, 551. The manager at the Ayeinm Mine in Ashanti paid seven pence per cord of firewood produced in 1903.

109. Report on the Mining Industry for the Year 1905, s. 1, PRO CO 98/14.

110. Coal cost around thirty shillings per ton in 1906. PRAAD ADM 5/1/15, 15.

111. Report on the Mining Industry for the Year 1905, s. 4, PRO CO 98/14.

112. Wills and Barrett, *Anglo-African Who's Who*, 195.

113. Wills and Barrett, *Anglo-African Who's Who*, 195.

114. "Abbontiakoon (Wassaw) Mines Limited," 675.

115. "Abbontiakoon (Wassaw) Mines Limited," 675.

116. Wills and Barrett, *Anglo-African Who's Who*, 195.

117. "The 'Lloyd' Copper Company, Ltd.," *Financial Times*, May 29, 1899, 3.

118. "Company Report of the Taquah Mining & Exploration," *Mining Magazine* 2 (1910): 155.

119. Guggisberg, *We Two in West Africa*, 1:107.

120. J. H. Curle, "West African Mines," *Mining Magazine* 1 (1909): 44.

121. "Abosso Gold," *Financial Times*, December 16, 1905, 2.

122. "Company Report of the Taquah Mining & Exploration," *Mining Magazine* 5 (1911): 75.

123. "Company Report of the Taquah Mining & Exploration," *Mining Magazine* 2 (1910): 155.

124. See separate liquidation notices by T. J. Foster for the Abosso Gold Mining Company Limited and the Taquah [*sic*] Mining and Exploration Company, Limited, *London Gazette*, March 20, 1923, 2195.

125. The Tarkwa and Abosso Company took over the Compagnie Minière de la Côte d'Or d'Afrique and the Compagnie Minière de la Côte d'Or Aboso in 1888. Dumett, *El Dorado in West Africa*, 107–8, table 4.1.

126. William F. Morecroft. "In the Matter of the Taquah and Abosso Gold Mining Company Limited," *London Gazette*, January 6, 1899, 101; Lewis Cappel, "The Taquah and Abosso Gold Mining Company Limited," *London Gazette*, January 16, 1900, 315.

127. Report on the Mining Industry for the Year 1905, s. 6, PRO CO 98/14.

128. Report on the Mining Industry for the Year 1905, s. 1.

129. Report on the Mining Industry for the Year 1905, s. 1.

130. "Taquah and Abosso Gold Mining Company (1900) Limited," *Economist*, December 16, 1905, 2037. Profits continued to fall. In 1907, 39,435 tons of ore were crushed and 11,055 tons of old tailings, through which 27,607 ounces of gold were retrieved, reaching an estimated value of £117,275. This added up to a decrease of £19,501 from the previous year. PRAAD ADM 5/1/16, 21.

131. "Taquah and Abosso Gold Mining Company (1900) Limited," 2037.

132. Report on the Mining Industry for the Year 1905, s. 5, PRO CO 98/14.

133. Report on the Mining Industry for the Year 1905, s. 5.

134. Report on the Mining Industry for the Year 1905, s. 5.

Chapter Two

1. "Mine Manager F. Stone to the Secretary for Mines," July 11, 1903, LMA MS14171, 12:555.

2. "Mine Manager F. Stone to the Secretary for Mines."

3. The term *Krumen* actually described a large ethnolinguistic family, which included the Bete, Dida, Grebo, and Wobe peoples, and other social groups.

4. For the full report, see "Report to the Directors of the Cote d'Or Company," February 14, 1896, Cade Papers.

5. PP 1889, No. 66 in [C 5620–24], 14.

6. Mines and Quarries: General Report and Statistics for 1904, PP 1906, [Cd. 2734, 2745, 2911], 599.

7. Dumett has written extensively about the role of tributary laborers in Gold Coast commerce. See "The Nzemans of Southwestern Ghana: Gold Miners, Rubber Traders, Loggers and Entrepreneurs" in *Ghana in Africa and the World: Essays in Honor of Adu Boahen*, ed. Toyin Falola (Trenton, NJ:, Africa World Press, 2003), 455–76. For the role of tributary men in the mechanized gold mines, see Dumett, *El Dorado*, 152–54.

8. Gareth Austin, *Labour, Land, and Capital in Ghana: From Slavery to Free Labour in Asante, 1807–1956* (Rochester, NY: University of Rochester Press, 2005), 102–4.

9. Burton and Cameron, *To the Gold Coast for Gold*, 2:333.

10. Burton and Cameron, *To the Gold Coast for Gold*, 2:333.

11. Burton and Cameron, *To the Gold Coast for Gold*.

12. PP 1889, No. 66 in [C 5620–24], 11.

13. PP 1889, No. 66 in [C 5620–24], 12.

14. Commander Rumsey to Colonial Secretary, September 4, 1882, PP 1883, Enclosure 1 in No. 19 in [C 3687], 54.

15. PP 1889, No. 66 in [C 5620–24], 12.

16. Crisp, *The Story of an African Working Class*, 20.

17. PRAAD ADM 12/3/1, 94.

18. Dumett, *El Dorado in West Africa*, 110.

19. Dumett, *El Dorado in West Africa*, 111.

20. Arthur Bowden of the Compagnie des Mines d'Or Abosso to the Colonial Secretary, July 11, 1883, PRAAD ADM 11/1/845, Political No. 18.

21. Arthur Bowden to the Colonial Secretary.

22. Arthur Bowden to the Colonial Secretary.

23. Arthur Bowden to the Colonial Secretary.

24. Arthur Bowden to the Colonial Secretary.

25. Arthur Bowden to the Colonial Secretary.

26. Burton and Cameron, *To the Gold Coast for Gold*, 2:112.

27. Burton and Cameron, *To the Gold Coast for Gold*, 2:112.

28. Isaacus Adzoxorno, "The Origin and Development of the Individual Contract of Employment in Ghanaian Labour Law" (paper delivered at the 1987 Conference of the African Studies Association of Australasia and the Pacific, Adelaide University, August, 1987), 27. It is also worth highlighting that the first men to sign individual contracts on the Gold Coast were migrant laborers from other parts of West Africa, such as Liberia. Adzoxorno, 112–13.

29. Compare table 2.1 with table 2.2.

30. Burton and Cameron, *To the Gold Coast for Gold*, 2:333.

31. Burton, letter to *Mining World*, March 17, 1883, 283.

32. Burton, letter to *Mining World*, March 17, 1883, 283.
33. Burton, letter to *Mining World*, March 17, 1883, 283.
34. Burton and Cameron, *To the Gold Coast for Gold*, 2:336–37.
35. Burton and Cameron, *To the Gold Coast for Gold*, 2:337.
36. Burton and Cameron, *To the Gold Coast for Gold*, 2:337.
37. Burton and Cameron, *To the Gold Coast for Gold*, 2:337.
38. Burton and Cameron, *To the Gold Coast for Gold*, 2:337.
39. Burton and Cameron, *To the Gold Coast for Gold*, 2:337.
40. Burton and Cameron, *To the Gold Coast for Gold*, 2:337.
41. Burton and Cameron, *To the Gold Coast for Gold*, 2:337.
42. David Kimble, *A Political History of Ghana: The Rise of Gold Coast Nationalism, 1850–1928* (Oxford: Oxford University Press, 1963), 24; Crisp, *The Story of an African Working Class*, 21–2, 26; Akurang-Parry, "'We Cast about for a Remedy': Chinese Labor and African Opposition in the Gold Coast, 1874–1914," *International Journal of African Historical Studies* 34 (2001): 365–84.
43. Akurang-Parry, "'We Cast about for a Remedy,'" 368.
44. Akurang-Parry, "'We Cast about for a Remedy,'" 368.
45. George G. Findlay, *The History of the Wesleyan Methodist Missionary Society* (London: Epworth Press, 1921), 4:150.
46. Findlay, *The History of the Wesleyan Methodist Missionary Society*, 4:150.
47. Was the expansive employment of foreign indentured laborers during the 1870s and 1880s an affront to surrounding populations, who also wanted to participate in the wage labor market, albeit at better wage rates and under more flexible conditions? Kwabena Akurang-Parry, for one, has shown that the importation of Chinese laborers did cause a political stir among African elites in colonial Ghana, who were discontented by the racist image of the African worker that this scheme underlined. See "'We Cast About for a Remedy.'"
48. Louis F. Gowan (February 1883), quoted in R. F. Burton, letter to the editor, *Mining World and Engineering Record*, April 28, 1883, 425.
49. Gowan, quoted in Burton, letter to the editor, *Mining World and Engineering Record*, April 28, 1883, 425.
50. J. D. Russell, "Pidgin-English amongst the Khroos," *Work and Workers in the Mission Field* (Wesleyan Methodist Missionary Society, 1909), 206.
51. Russell, "Pidgin-English amongst the Khroos."
52. Russell, "Pidgin-English amongst the Khroos."
53. Although these names could stick with laborers for quite some time, some European employers preferred to give their men new ones. The Methodist missionary Russell explained: "As a rule, these boys rejoice in the weirdest names imaginable. . . . Soon after they came here I changed their names and Black-Man-Trouble is now known as Tom, while John

Railway-Garden has become Sam." Russell, "Pidgin-English amongst the Khroos."

54. Migeod to Governor Rodgers, April 28, 1906, Cambridge University Library (hereafter CUL) RCMS 139/10/3, no. 661/06.

55. Generally, Kru laborers were informed about the condition of work at their future workplace only in vague terms before leaving their homes, but a drastic departure from their expectations could incite collective violent resistance. According to one anecdote: "A curious incident took place in 1858. The French ship, *Regina Coeli*, arrived on the Kru Coast, and the Captain treated with Kru chiefs for men to be shipped as laborers; the men supposed that they were shipped for a trip along the west coast, as usual, to serve as seamen; learning, however, that their destination was the West Indies, they became alarmed and believed that they were to be sold into slavery; the captain was still on shore, treating with the chiefs; the men mutinied, seized the ship, and killed all the white crew except the doctor; they then returned to shore and left the ship without a crew; had she not been noticed by a passing English steamer, she would no doubt have been wrecked; she was taken into a Liberian port." Frederick Starr, *Liberia: Description, History, Problems* (Chicago, 1913), 91.

56. Migeod to Governor Rodgers, April 28, 1906.

57. Migeod to Governor Rodgers.

58. Migeod to Governor Rodgers.

59. It is relevant to debates on "cheap-labor theory that employers in Wassa perceived the recurring patterns of oscillatory migration that Liberian contract men followed at the end of their twelve-month contract as a problem. For instance, Burton and Cameron were clearly annoyed when they argued that the "men's services are lost just as they are becoming valuable." *To the Gold Coast for Gold*, 2:334. Also, according to one administrator, "Another disadvantage to progress is the necessity of getting fresh kroomen; they will only engage for a year at a time, and nothing will induce them to remain longer, and the consequence is that about the time they are really becoming skilled and useful labourers, their period of service is up and they themselves off to their homes." Fifth Monthly Report on the Tarquah Gold Mining District, September 21, 1882." PP 1883, Enclosure in No. 27 in [C 3687], 73.

60. Fifth Monthly Report on the Tarquah Gold Mining District.

61. Brown, "*We Were All Slaves,*" 97.

62. Brown, "*We Were All Slaves,*" 78.

63. Lentz, *Ethnicity and the Making of History*, ch. 5.

64. Transport Circular, December 2, 1904, CUL RCMS 139/11/8.

65. "A. D. Wilson—Native Interpreter," June 7, 1902, LMA MS14171, vol. 1, D. 366.

66. Commissioner of Labour J. R. Dickinson (1939), quoted in Crisp, *The Story of an African Working Class*, 69.

67. Starting in the late nineteenth century, "unskilled" Liberian laborers began to travel further along the coast of West Africa to find work in growing numbers. Jane Martin, "Krumen 'Down the Coast': Liberian Migrants on the West African Coast in the 19th and early 20th Centuries," *International Journal of African Historical Studies* 18, no. 3 (1985): 401–23; Diane Frost, *Work and Community among West African Migrant Workers since the Nineteenth Century* (Liverpool: Liverpool University Press, 1999), 20.

68. William Allen and T. R. H. Thomson, *A Narrative of the Expedition Sent by Her Majesty's Government to the River Niger in 1841* (London: Frank Cass & Co Ltd., 1841), 1:77.

69. Letters from Colonial Agents, 1833–40, 51–54, Liberian Government Archives I for 1828–1911, Archives Online at Indiana University.

70. Letters from Colonial Agents, 1833–40, 51–54.

71. Letters from Colonial Agents, 1833–40, 51–54.

72. Allen, *A Narrative of the Expedition*, 1:494.

73. Report by C. W. Mann to Directors of the Ashanti Goldfields Corporation, January 10, 1905, 11:48–54, LMA MS14171.

74. Report on the Transport Department for the Year 1903, s. 5, RHO MS 7ss.14 s.10.

75. Report on the Transport Department for the Year 1903.

76. Guggisberg, *We Two in West Africa*, 1:122.

77. Guggisberg, *We Two in West Africa*, 1:121.

78. W. J. Loring, "The West African Outlook," *Kalgoorlie Western Argus*, August 3, 1909, 8.

79. Loring, "The West African Outlook," 8.

80. Quoted in Guggisberg, *We Two in West Africa*, 1:8.

81. Quoted in Guggisberg, *We Two in West Africa*, 1:8.

82. Quoted in Guggisberg, *We Two in West Africa*, 1:8.

83. Quoted in Guggisberg, *We Two in West Africa*, 1:8.

84. "The Gold Fields of the West Coast of Africa," *Australian Town and Country Journal*, March 1, 1902, 25.

85. "The Gold Fields of the West Coast of Africa," 25.

86. Loring, "The West African Outlook"

87. "The Gold Fields of the West Coast of Africa," 25.

88. An excerpt from a report by District Commissioner Cuscaden captures the frequent, unexpected disruptions to white labor productivity in Tarkwa: "Mr. Criswick, manager of Gold Coast Company, went home sick on the 25th March. The doctor of the French company died this month. Another European belonging to same company was sent down, and died on board. Captain Burton sailed for England beginning of the month. He came as far as Tomento, but became ill and had to go back. The manager

[of] 'Effuenta Company' went home sick at the same time. Commander Cameron sailed for England on the 23rd instant." Cuscaden, First Monthly Report on Tarquah Gold Mining District, April 30 1882, Enclosure in No. 3 in [C 3687], 10, PP 1883.

89. "The Gold Fields of the West Coast of Africa," 25.

90. "West African Mining," *Mining Magazine* (July–December 1910): 3:322.

91. Guggisberg, *We Two in West Africa*, 1:127.

92. "Fanti Consolidated Mines, Limited."

93. "The Gold Fields of the West Coast of Africa," 25.

94. "Fanti Consolidated Mines, Limited."

95. "Fanti Consolidated Mines, Limited."

96. "Fanti Consolidated Mines, Limited."

97. "The Gold Fields of the West Coast of Africa," 25.

98. "Fanti Consolidated Mines, Limited": 249.

99. Transport Circular No. 4, CUL RCMS 139/11/8.

100. Allen, *A Narrative of the Expedition*, 1:122.

101. Allen, *A Narrative of the Expedition*, 1:123.

102. Allen, *A Narrative of the Expedition*, 1:123.

103. Allen, *A Narrative of the Expedition*, 1:123.

104. Allen, *A Narrative of the Expedition*, 1:123.

105. Jean Allman, "Let Your Fashion Be in Line with Our Ghanaian Costume," in *Nation, Gender, and the Politics of Clothing in Nkrumah's Ghana* (Bloomington: University of Indiana Press, 2004), 147.

106. Lentz, *Ethnicity and the Making of History*, 7.

107. Lentz, *Ethnicity*, 34.

108. Lentz, *Ethnicity*, 2–3. Lentz, *Land, Mobility, and Belonging in West Africa* (Bloomington: Indiana University Press, 2013).

109. Lentz, *Ethnicity*, 34.

110. Report on the Transport Department for the Year 1903, s. 5, RHO MS 7ss.14 s.10.

111. Report on the Transport Department for the Year 1903.

112. Carola Lentz and Veit Erlmann, "A Working Class in Formation? Economic Crisis and Strategies of Survival among Dagara Mine Workers in Ghana," *Cahiers d'Études africaines* 29 (1989): 102.

113. Diane Frost describes how around this period, Kru seamen working under contract each deposited one shilling per working day with their headmen or gang leaders, who in turn would deposit the money with a bank. This communal fund, the "Kru Fund," was intended to go only toward supporting colleagues during times of hardship, including illness or death, or to educate the offspring of poor laborers. Frost, *Work and Community among West African Migrant Workers*, 42.

114. Letter No. 3, CUL RCMS 139/16.

115. Brown, *"We Were All Slaves,"* 58.

116. Russell, "Pidgin-English Amongst the Khroos," 206.

117. Paul Jenkins, "The Anglican Church in Ghana, 1905–24 (I)," *Transactions of the Historical Society of Ghana*, 1 (June 1974): 25–26. The intersection of labor and religion certainly deserves increased historical investigation. However, what is in under dispute here is the extent of control that gang leaders had over their recruits.

118. Frankema and van Waijenburg, "Structural Impediments to African Growth?"

119. George Bottoms to Martin Bottoms, November 4, 1906, cited in Dorward, "'Nigger Driver Brothers,'" 205.

120. Gaunt, *Alone in West Africa*, 325.

121. Memorandum on the Concessions Labour Bill, enclosed in confidential dispatch, chief justice to chief officer of the Transport Department, No. Case 714, March 25 1904, CUL RCMS 139/12/11. This behavior may have been further encouraged by the scarcity of cash in the southern Gold Coast. Austin has mentioned that because cash was scarce, and because of the time and effort that needed to be invested in securing a loan, a decent profit could be made from lending cash at a sufficient interest rate. *Labor, Land, and Capital in Ghana*, 369.

122. Memorandum on the Concessions Labour Bill, CUL RCMS 139/12/11.

123. Letters from Colonial Agents, 1833–1840, 51–54.

124. Memorandum on the Concessions Labour Bill, CUL RCMS 139/12/11.

125. Hill, *Development Economics on Trial*, ch. 8.

126. Memorandum on the Concessions Labour Bill, CUL RCMS 139/12/11.

127. Acting Governor Herbert Bryan to Secretary of State, May 21, 1906, No. 24, "Coolie Labour," PP 1906 (357), 34.

128. Ravi Ahuja, "Mobility and Containment: Voyages of South Asian Seamen, c. 1900 to 1960," in *Coolies, Capital and Colonialism: Studies in Indian Labour History*, ed. Rana Behal and Marcel van der Linden (Cambridge: Cambridge University Press, 2007), 135.

129. Adzoxorno, "The Origin and Development of the Individual Contract," 10.

130. Carnarvon to Governor Strahan, August, 21, 1874, reprinted in George Edgar Metcalfe, ed., *Great Britain and Ghana: Documents of Ghana History 1807–1957* (London: Thomas Nelson, 1964), 373.

131. Chada has previously demonstrated how the Master and Servant Act of 1893 protected workers in the Gold Coast by upholding contractural agreements. G. T. Z. Chada, "Labor Protests, Group Consciousness and Trade Unionism in West Africa: The Radical Railway Workers of

Colonial Ghana" (PhD diss., University of Toronto, 1981). At the same time, Akurang-Parry has shown that later amendments were geared toward increasing the labor supply in the Gold Coast by restricting outside migration. Kwabena Opare Akurang-Parry, "Missy Queen in Her Palaver Says De Gole Cosse Slaves Is Free: The British Abolition of Slavery/Pawnship and Colonial Labor Recruitment in the Gold Coast [Southern Ghana], 1874– Ca. 1940" (PhD diss., York University, 1999).

132. Master and Servant Act of the Gold Coast Colony of 1893.

133. Adzoxorno, "The Origin and Development of the Individual Contract," 27.

134. J. D. Jally v. Englishman, Jacob, Serge, Criminal and Civil Record Book, District Commissioner's Court, October 28 and November 1, 1909, 708–9, 719, PRAAD ADM 27/4/49.

135. Superintendent J. A. Strong v. Kwesi K., 231, PRAAD ADM 27/4/45.

136. "Fanti" or "Fantee" was a term Europeans often used to refer to all social groups from the coastal region of the Gold Coast Colony and protectorate.

137. Letter no. 3, February 3, 1910, RCMS 139/12/52.

138. Letter no. 3. Unfortunately, there is little detailed information in the sources to explain how this risk was contained.

139. Rex v. Joe Mendi, Criminal and Civil Record Book, District Commissioner's Court, May 7, 1904, 307–8, PRAAD ADM 27/4/44.

140. Inspector General of Police v. Bukai Bazahim, Criminal and Civil Record Book, District Commissioner's Court, 1914, 306, PRAAD ADM 27/4/54.

141. Inspector General of Police v. Bukai Bazahim.

142. Rex v. 12 Illasi; 75 Salakor; 38 Abbey; 22 Adessoman; 34 Hadji, Criminal and Civil Record Book, District Commissioner's Court, September 21, 1904, 599–601, PRAAD ADM 27/4/43.

143. Rex v. 12 Illasi; 75 Salakor; 38 Abbey; 22 Adessoman; 34 Hadji.

144. See also Rex pro Kofi H. v. Asamoah, Criminal and Civil Record Book, District Commissioner's Court, March 6, 1893, 426, PRAAD ADM 27/4/40.

145. Rex v. 12 Illasi; 75 Salakor; 38 Abbey; 22 Adessoman 34 Hadji.

146. Rex pro Superintendent Phillips v. Bokai, Criminal and Civil Record Book, District Commissioner's Court, 1903, 89, PRAAD ADM 27/4/43.

147. Rex pro Superintendent Phillips v. Bokai.

148. Rex pro Superintendent Phillips v. Bokai.

149. Master and Servant Act of the Gold Coast Colony of 1893. Ordinance No. 8, PP, Gold Coast Laws and Regulations Affecting Land, Mines and Labor of Natives (1901), 42.

150. "Rates of Wages and Cost of Living," Gold Coast Departmental Report for 1904, PRAAD ADM 5/1/13, 49.

151. Report on the Transport Department for the Year 1903, s.4, RHO MS 7ss.14 s.10.

152. Report on the Transport Department for the Year 1908, s. 5, PRO CO 98/16.

153. Report on the Transport Department for the Year 1903, s. 4, RHO MS 7ss.14 s.10.

154. Adzoxorno, "The Origin and Development of the Individual Contract," 3.

155. These letters are kept as part of the private papers of F. W. H. Migeod.

156. Conah Mandorrah to Gin Bottle, June 22, 1912, CUL RCMS 139/16.

157. Cona Mandorrah to Gin Bottle.

158. Conah Mandorrah to Gin Bottle.

159. Conah Mandorrah to Gin Bottle.

160. Conah Mandorrah to Gin Bottle.

161. Conah Mandorrah to Gin Bottle.

162. Family to Lacton, 9 February 1919, CUL RCMS 139/16.

163. Family to Lacton.

164. Family to Lacton.

165. Family to Lacton.

166. Family to Lacton.

167. Alimendi to Chief Momo Gbainya, [n.d.], CUL RCMS 139/16.

168. For rural-based economies in colonial West Africa, Polly Hill has previously described how the poorest members of rural communities were the least likely to migrate to new areas. She has also convincingly argued that the poorest members of such communities were the least likely to get access to credit due to their socioeconomic standing within the community. *Development Economics on Trial*, ch. 8.

169. Alimendi to Momo Gbainya.

170. Joseph Bomah to his father, 1919, CUL RCMS 139/16.

171. Joseph Bomah to his father.

172. Allen, *A Narrative of the Expedition*, 1:122–23.

173. Allen, *A Narrative of the Expedition*, 1:123.

Chapter Three

1. For the full report, see Report to the Directors of the Côte d'Or Company, February 14, 1896, Cade Papers.

2. Report to the Directors of the Côte d'Or Company.

3. Starr, *Liberia*, 112–13.

4. Starr, *Liberia*, 113. The southeast region of Liberia, as far east as the San Pedro River now lying some fifty miles from the modern border of the Ivory Coast, had originally come into the possession of the Liberia government in 1857. Officials still held the deeds of purchase and treaties to prove it.

5. Starr, *Liberia*, 114–15.

6. Starr, *Liberia*, 114–15.

7. Starr, *Liberia*, 114–15.

8. Raymond Leslie Buell, *The Native Problem in Africa* (New York: Macmillan, 1926), 2:777, 789. Diane Frost puts forward that labor was Liberia's main export in the nineteenth century due to restricted economic development. The tax that was levied on migrants going to work for wages elsewhere in the twentieth century had the effect of restricting the overall number of available contract laborers along the West African coast. *Work and Community among West African Migrant Workers since the Nineteenth Century*, 10.

9. Christine Behrens, *Le Croumen de la côte occidentale de l'Afrique* (Bordeaux: Ministère de l'Éducation Nationale, 1974), 65–82.

10. The Liberian president declared, "I trust you will see the importance of discouraging any proposal looking to the removal of labor out of the country." Message of the President, in Starr, *Liberia*, 113.

11. The price of coffee fell from around twenty-five cents a pound at the height of its popularity on the European market in the late nineteenth century to about eight cents a pounds after 1900. Starr, *Liberia*, 136.

12. Gold Coast Departmental Report for the Year 1903, 50.

13. Bullen therefore requested that the government transport department reduce the one-pound deposit that the department requested for the provision per individual gang of contract laborers. James Bullen the Hon. Secretary of the Mine Managers' Association to Colonial Office, Abbontiakoon, May 11, 1902, CUL RCMS 139/12/25.

14. Gold Coast Departmental Report for the Year 1903.

15. Migrants from French colonies, such as Mali and Burkina Faso, did not seek out jobs in the Gold Coast mines in large numbers until after World War I. Report on the Mining Industry for the Year 1918, RHO MS 7ss 13 s.4.

16. Gold Coast Departmental Report for the Year 1903, 49.

17. Gold Coast Departmental Report for the Year 1903.

18. Gold Coast Departmental Report for the Year 1903.

19. Gold Coast Departmental Report for the Year 1903, 50.

20. Report on the Transport Department for the Year 1903, s. 6, RHO MS 7ss.14 s.10.

21. Report on the Transport Department for the Year 1903, s. 6.

22. Colonial observers, however, hoped that with the completion of the government railway, the dependence on porterage would decline and that these men and women would eventually be obliged to take on work elsewhere, including the mining sector. Report on the Transport Department for the Year 1903, s. 6, RHO MS 7ss.14 s.10.

23. Report on the Transport Department for the Year 1903, s. 6.

24. For example, in 1906 the chief officer of the government transport department expressed some hostility toward casual laborers: "The Mendes and Yorubas come to this colony on their own initiative and take up such work as offers. They stay with an employer so long as it suits them and when the inclination takes them to make a change they merely await pay day to leave." Migeod to Governor Rodgers, April 28, 1906.

25. Spelled "Quow" in the source material.

26. Quow Hammah v. Mahoma Nuhu, Criminal and Civil Record Book, District Commissioner's Court, 1894, 39–42, PRAAD ADM 27/4/41.

27. Quow Hammah v. Mahoma Nuhu.

28. Quow Hammah v. Mahoma Nuhu.

29. Farewell Circular to the Staff of the Gold Coast Transport Department on its Abolition, June 23, 1919, CUL RCMS 139/11/37.

30. Farewell Circular to the Staff of the Gold Coast Transport Department.

31. Farewell Circular to the Staff of the Gold Coast Transport Department.

32. Assistant Transport Officer to Chief Officer of the Transport Department, 1903, CUL RCMS 139/6.

33. Guggisberg, *We Two in West Africa*, 1:138.

34. Guggisberg, *We Two in West Africa*, 1:138.

35. Guggisberg, *We Two in West Africa*, 1:138.

36. Guggisberg, *We Two in West Africa*, 1:138.

37. Guggisberg, *We Two in West Africa*, 1:138.

38. Guggisberg, *We Two in West Africa*, 1:138.

39. Guggisberg, *We Two in West Africa*, 1:138.

40. Farewell Circular to the Staff of the Gold Coast Transport Department, June 23, 1919.

41. Eamonson, August 23, 1902, LMA MS14171, vol. 1, s. 117.

42. Eamonson, LMA MS14171, vol. 1, s. 117.

43. N. J. Suckling to the Secretary of the Ashanti Gold Fields Corporation, June 21, 1902, LMA MS14171, vol. 1, s. 65.

44. N. J. Suckling to the Secretary of the Ashanti Gold Fields Corporation, June 21, 1902.

45. N. J. Suckling to the Secretary of the Ashanti Gold Fields Corporation, June 21, 1902.

46. Mr. Eamonson, LMA MS14171, vol. 1, D. 393.

47. Mr. Eamonson, LMA MS14171, vol. 1, D. 393.

48. N. J. Suckling to the Secretary of the Ashanti Goldfields Corporation," August 23, 1902, LMA MS14171, vol. 1, s. 117.

49. N. J. Suckling to the Secretary of the Ashanti Goldfields Corporation," August 23, 1902.

50. N. J. Suckling to the Secretary of the Ashanti Goldfields Corporation, August 23, 1902.

51. N. J. Suckling to the Secretary of the Ashanti Goldfields Corporation, August 23, 1902.

52. N. J. Suckling to the Secretary of the Ashanti Goldfields Corporation, August 23, 1902.

53. Alan Jeeves, *Migrant Labour in South Africa's Mining Economy: The Struggle for the Gold Mines' Labor Supply, 1890–1920* (Kingston, ON: McGill-Queen's University Press, 1985).

54. Lawrance, Osborn, and Roberts, introduction to *Intermediaries, Interpreters, and Clerks*, 4.

55. Jeeves, *Migrant Labour in South Africa's Mining Economy*, 97.

56. Criminal and Civil Record Book, District Commissioner's Court, 1909, 762–63, PRAAD ADM 27/4/48.

57. Rex v. J. Fry, Criminal and Civil Record Book, District Commissioner's Court, May 12, 1904, 329–30, PRAAD ADM 27/4/43.

58. Rex v. J. Fry.

59. Rex v. J. Fry.

60. Shandi to Chief Officer of the Transport Department, Sekondi, February 13, 1910, CUL RCMS 139/16.

61. Shandi to Chief Officer of the Transport Department.

62. Inspector General of Police v. Bonsa Kwamin, Criminal and Civil Record Book, District Commissioners Court, March 1919–July 1920, 223, 225, PRAAD ADM 27/4/61.

63. See for example, Inspector General of Police v. Seiden Wangara, 276–77, PRAAD ADM 27/4/60.

64. Inspector General of Police v. Seiden Wangara.

65. Inspector General of Police v. Seiden Wangara.

66. Inspector General of Police v. Seiden Wangara.

67. Inspector General of Police v. Mana Wangara.

68. Mine Supplier Taquah Mine v. Rodson, Criminal and Civil Record Book, District Commissioner's Court, 1915, 46–57, PRAAD ADM 27/4/57.

69. Mine Supplier Taquah Mine v. Rodson.

70. Mine Supplier Taquah Mine v. Rodson.

71. Mine Supplier Taquah Mine v. Rodson.

72. Inspector General of Police v. Seiden Wangara, 276–77, PRAAD ADM 27/4/60.

73. Inspector General of Police v. Seiden Wangara.

74. Inspector General of Police v. Seiden Wangara.

Chapter Four

1. Although its membership was much broader, the correspondence of the Mine Managers' Association always came from managers on these two properties, according to the scattered source material gathered by the author. Its secretaries and presidents were also continuously chosen from within the rank of upper-level managers on these two concessions.

2. Crisp, *The Story of an African Working Class*, 27.

3. James Bullen the Hon. Secretary of the Mine Managers' Association to Colonial Office, Abbontiakoon, May 11, 1902, CUL RCMS 139/12/25.

4. "White Labour on the Rand," *Economist*, October 24, 1903, 1798.

5. "White Labour on the Rand." After the war, the South African government took steps to provide support to unemployed British ex-servicemen who had remained in the colony. The British High Commissioner, Alfred Milner, developed a scheme together with the mines to employ them in unskilled mining positions, though at wage rates much higher than those Africans had received for the same work. Whereas the ex-servicemen deemed the wages to be too low, the work to be too severe, and the lifestyle in the mining centers to be substandard, in contrast many Afrikaans-speakers grasped the opportunity to work their way up into skilled mining positions. This experiment however, never was considered a long-term solution. Dagmar Engelken, "A White Man's Country: The Chinese Labour Controversy in the Transvaal," in *Wages of Whiteness and Racist Symbolic Capital*, ed. Wulf D. Hund, Jeremy Krikler, and David R. Roediger (Berlin: LIT Verlag, 2010), 167.

6. Engelken, "A White Man's Country," 166–67.

7. John Hamill, *The Strange Career of Mr. Hoover under Two Flags* (New York: J. Day Co., 1931), 152. In her 2010 publication, Dagmar Engelken shows how the "Chinese labor question" drove the formation of political alliances and parties in the aftermath of the First South African War. "A White Man's Country."

8. Though in the end he was not able to convince mine managers to adapt a similar strategy, Creswell continued the fight to hire more unemployed white men. He went on to join the South African Labor Party, eventually becoming its leader in the South Africa parliament. Engelken, "A White Man's Country," 168–69.

9. Skilled white miners were also opposed to Creswell's suggestions for the widespread employment of white workers in unskilled positions because: (1) their presence would endanger the high wages that other whites had had access to until this point; (2) skilled white workers

consequently would be forced into supervisory positions that exposed them to harsh, disease-inducing environments; (3) they felt it would be generally disruptive to the racial solidarity that existed among whites at the mines; (4) they opposed Creswell's related proposal to import Europeans as part of this scheme; and (5) they feared being replaced by unskilled white workers over time.

10. "Abbontiakoon (Wassaw) Mines Limited," *Economist*, November 15, 1902, 1773.

11. "Abbontiakoon (Wassaw) Mines Limited."

12. "Abbontiakoon (Wassaw) Mines Limited."

13. "Wassau (Gold Coast) Mining Company, Limited," *Economist*, March 23, 1901, 452.

14. "Wassau (Gold Coast) Mining Company, Limited," 452.

15. "Wassau (Gold Coast) Mining Company, Limited," 452. Starting in the 1890s, a private centralized labor bureau was set up in Johannesburg to regulate all recruitment activities, in order to restrict competition from employers in other sectors and to limit competition within the gold-mining industry itself.

16. "Gold Coast Investment Company Limited," 452. Jones and his associates called for government-supported commercial cotton production overseas in response to a worldwide cotton shortage around this time. However, his proposal for a monopoly on labor was only relevant prior to 1902, when the government actually considered the economic potential of plantations managed by Europeans. In the end the government concluded that any agricultural development would have to be led by a multitude of small African producers. Dumett, "Obstacles to Government-Assisted Agricultural Development in West Africa," 159–61. Jones, who was president of the Chamber of Commerce in Liverpool and nicknamed "Uncrowned King of West Africa," had expansive investments in West Africa. For biographical information, see Alan Hay Milne, *Sir Alfred Lewis Jones K. C. M. G.: A Story of Energy and Success* (Liverpool: Henry Young & Sons, 1914).

17. "Wassau (Gold Coast) Mining Company, Limited," 452.

18. "Wassau (Gold Coast) Mining Company, Limited," 452.

19. "Wassau (Gold Coast) Mining Company, Limited," 452.

20. Jeff Crisp, "The Labour Question on the Gold Coast," in *Proletarianisation in the Third World: Studies in the Creation of a Labour Force under Dependent Capitalism*, ed. Henry Finch and Barry Munslow (London: Routledge, 1984), 31.

21. Bullen to Colonial Office, CUL RCMS 139/12/25.

22. Bullen to Colonial Office.

23. Bullen to Colonial Office.

24. Bullen to Colonial Office.

25. Bullen to Colonial Office. For example, the record book would list the arrival and departure dates of white workers, as well as whether or not they were on contract, in order to facilitate accounting.

26. Crisp, "The Labour Question on the Gold Coast."

27. In 1905, this organization was succeeded by the West African Chamber of Mines.

28. Herbert W. L. Way, *Round the World for Gold: A Search for Minerals from Kansas to Cathay* (London: Sampson Low, Marston & Company, Ltd., 1912), 332. The Concessions Ordinance of 1900 allowed the chiefs to dispose of communal and tribal lands without the intervention of the government. It also safeguarded the interests of African owners by empowering the Supreme Court to invalidate concessions that had been unfairly acquired.

29. Report on the Transport Department for the Year 1902 (1), s. 6, RHO MS 7ss.14 s.10.

30. In 1906, the Government Transport Office also established a station in the Northern Territories.

31. Initially, laborers were hired mainly to load trains at Sekondi. At Dunkwa and Obuase, the supplies were collected from the trains and distributed by carriers. In addition, the department engaged porters to work along its mail route through Obuase and Kumase. "Report for the Gold Coast Colony for the Year 1906, 49–50."

32. Extract from His Excellency's Answer to the Proposals Submitted by the Mine Managers' Association, CUL RCMS 139/12/25.

33. Report on the Transport Department for the Year 1908, s. 7, PRO CO 98/16.

34. Extract from His Excellency's Answer to the Proposals Submitted by the Mine Managers' Association.

35. Chief Officer of the Transport Department to Manager Cinnamon Bippo, CUL RCMS, 139/6. After 1904, the Mine Managers' Association continued to request that the Government Transport Office be replaced by a labor bureau.

36. Report on the Transport Department for the Year 1903, s. 4, RHO MS 7ss.14 s.10.

37. Enclosure in Report on the Transport Department for the Year 1903.

38. Report on the Transport Department for the Year 1904, s. 1, PRO CO 98/14.

39. Report on the Transport Department for the Year 1905, s. 4, PRO CO 98/14.

40. Report on the Transport Department for the Year 1907, s. 1, PRO CO 98/16.

41. Migeod to Governor Rodgers, April 28, 1906.

42. Chief Officer of the Transport Department to the Assistant Transport Officer at Obuase, CUL RCMS 139/6/19.

43. Chief Officer of the Transport Department to the Assistant Transport Officer at Obuase.
44. Migeod to Governor Rodgers, April 28, 1906.
45. Hiring Regulations, CUL RCMS 139/6/31.
46. Transport Circular No. 3, July 11, 1903, Official Papers on Labour Policy, CUL RCMS 139/12/10.
47. Transport Circular No. 3, July 11, 1903.
48. Transport Circular No. 3, Official Papers on Labour Policy, July 11, 1903.
49. Report on the Transport Department for the Year 1903, s. 5, RHO MS 7ss.14 s.10.
50. Report on the Transport Department for the Year 1903, s. 5.
51. Chief Officer of the Transport Department to the Assistant Transport Officer at Obuase, Manifold Book, July–August 1903, CUL RCMS 139/6/26.
52. Transport Circular No. 3, Official Papers on Labour Policy, July 11, 1903, CUL RCMS 139/12/10.
53. Transport Circular No. 3, Official Papers on Labour Policy, July 11, 1903, CUL RCMS 139/12/10.
54. Transport Circular No. 3, July 11, 1903.
55. Report on the Transport Department for the Year 1903, s. 4, RHO MS 7ss.14 s.10. See also CUL RCMS 139/12/10.
56. Transport Circular No. 3, Official Papers on Labour Policy, July 11, 1903, CUL RCMS 139/12/10.
57. Transport Circular No. 3, Official Papers on Labour Policy, July 11, 1903, CUL RCMS 139/12/10.
58. Transport Clerk Atebubu A Edward Ammah to Chief Officer of the Transport Department, August 6, 1918, CUL RCMS 139/16.
59. Transport Clerk Atebubu to Chief Officer of the Transport Department.
60. Transport Clerk Atebubu to Chief Officer of the Transport Department.
61. Transport Clerk Atebubu to Chief Officer of the Transport Department.
62. Karthar to Chief Officer of the Transport Department, CUL RCMS 139/16.
63. The Prestea mines gained control of the properties of the Appantoo Mining Company, a reconstruction of the Gie Appantoo and Essaman Gold Mining Company, in 1893. See table 4.1. in Dumett, *El Dorado in West Africa*, 107–8.
64. Effuenta (Wassaw) Mines Limited to WH Migeod, 5 August 1903, CUL RCMS 139/12/31.

65. Most likely due to the type of communication, laborers were not reprimanded in any of the surviving letters held in the archive. Instead, the letters mostly showed Migeod warning employers that a bad reputation would have a huge impact on their company's recruitment efforts. In a dispute with the manager of the Tamsu mine, Migeod wrote, "I should much regret that any misunderstanding should arise between yourself and the labourers, but should like to see the work carried out in harmony and with good feeling. I trust therefore you may see your way to removing any misapprehension that exist in the labourers' minds, failing which they have informed me they will all leave the mine. . . . This would be a regrettable incident. It would be exceedingly difficult to find other men . . . and as desertion after the first few days rarely occurs without a solid grievance, since the men know they are liable [to] imprisonment in their case if not a good one. If their case should prove a good one, they may at least claim full pay for every day they have been at the mine." F. W. H. Migeod to the General Manager of W. W. A. M Tansu, January 5, 1903, CUL RCMS 139/4.

66. Headman Charles Kofi to the Pay Master, 1906, CUL RCMS 139/16.

67. Headman Charles Kofi to the Pay Master, 1906.

68. Headman Charles Kofi to the Pay Master, 1906.

69. Chief Officer of the Transport Department to General Manager W.W.A.M Tamsu, January 5, 1903, letter books, CUL RCMS 139/4.

70. Transport Circular No. 3, Official Papers on Labour Policy, July 11, 1903, CUL RCMS 139/12/10.

71. Transport Circular No. 3, Official Papers on Labour Policy, July 11, 1903, CUL RCMS 139/12/10.

72. Chief Officer of the Transport Department to General Manager E. Axim, CUL RCMS 139/4. It has elsewhere been stated that Migeod was in favor of the managers' prerogative to refuse subsistence to men on a Sunday, though there is little evidence for such a claim. L. H. Gann and Peter Duignan, *The Rulers of British Africa, 1870–1914* (Stanford: Stanford University Press, 1978), 273.

73. Anfargah Mutiny, October 9, 1903, CUL RCMS 139/4/30.

74. Anfargah Mutiny, October 9, 1903, CUL RCMS 139/4/30.

75. Manifold Book, July–August 1903, CUL RCMS 139/6/26. In this case it is not entirely clear whether the men stood to lose any money at all.

76. Manifold Book, July–August 1903.

77. Manifold Book, July–August 1903.

78. Chief Officer of the Transport Department to Colonial Secretary, December 15, 1903, letter books, CUL RCMS 139/4.

79. Chief Officer of the Transport Department to Colonial Secretary, December 15, 1903.

80. Chief Officer of the Transport Department to Colonial Secretary, December 15, 1903.

81. Lawrance, Osborn, and Roberts, introduction to *Intermediaries, Interpreters, and Clerks*.

82. Chief Officer of the Transport Department to Colonial Secretary, December 15, 1903.

83. Chief Officer of the Transport Department to Colonial Secretary, December 15, 1903.

84. Thora Williamson, *Gold Coast Diaries: Chronicles of Political Officers in West Africa, 1900–1919*, ed. Anthony Kirk-Greene (London: Radcliffe Press, 2000).

85. Gerhard Maier, *African Dinosaurs Unearthed: The Tendaguru Expeditions (Life of the Past)* (Bloomington: Indiana University Press, 2003), 165.

86. See Great Britain Colonial Office, *Colonial Office List for 1920* (London, 1920), 700.

87. Gann and Duignan, *The Rulers of British Africa*, 273n.

88. Royal Anthropological Institute of Great Britain and Ireland, *Man, a Monthly Record of Anthropological Science Nos. 1–138* (London, 1917).

89. George Gaylord Simpson, *Simple Curiosity: Letters from George Gaylord Simpson to His Family, 1921–1970* (Berkeley: University of California Press, 1987), 75.

90. Simpson, *Simple Curiosity*, 75.

91. Report on the Transport Department for the Year 1902 (1) s. 3, RHO MS 7ss.14 s.10.

92. Report on the Transport Department for the Year 1902 (1) s. 3, RHO MS 7ss.14 s.10.

93. Following several failed attempts to stop the mining companies from competing for labor on the Rand, in 1895 the Chamber of Mines persuaded the Volksraad to enact a pass law in certain so-called labor districts that would feed the gold mines. Jeeves, *Migrant Labour in South Africa's Mining Economy*, 42.

94. Raymond Dumett and Marion Johnson, "Britain and the Suppression of Slavery in the Gold Coast Colony," in *The End of Slavery in Africa*, ed. Suzanne Miers and Richard Roberts (Madison: University of Wisconsin Press, 1988), 92; Dumett, *El Dorado in West Africa*, 225.

95. Memorandum on the Concessions Labour Bill, CUL RCMS 139/12/11.

96. Memorandum on the Concessions Labour Bill.

97. Memorandum on the Concessions Labour Bill.

98. "Extract from His Excellency's Answer to the Proposals submitted by the Mine Managers' Association," response to James Bullen to Colonial Office, May 11, 1902, CUL RCMS 139/12/25.

99. "Extract from His Excellency's Answer."

100. Report on the Transport Department for the Year 1908, s. 4, PRO CO 98/16.

101. Report on the Transport Department for the Year 1908, s. 4.

102. (Counterfoil) Certificate of Registration of a Native in Connection with the Concessions Labour Ordinance, CUL RCMS 139/12/40.

103. (Counterfoil) Certificate of Registration.

104. Half a decade later, in 1908, the government transport department took registration even further by implementing a policy of not just recording the enlistment of carriers on an individual basis with their thumbprints but also compelling them to wear labels, each showing a unique number, while at work. An official report was explicit about how European supervisors benefited from this form of identification: "For the first time, tallies have been issued to the permanent carriers belonging to the Transport Department. As every man is known to the officers of the department, they are of no special value from a departmental point of view, though officers travelling with these carriers find the fact that every man bears a number of considerable assistance for identification. Seeing that nearly every man bears two names, his own personal name which his friends use, and his adopted name taken only for dealings with Europeans, which he not infrequently forgets as he has probably changed it for every employer who he has served, the use of a tally becomes apparent." Registration was primarily meant to halt desertion, but Migeod attempted to give further authority to the experiment by claiming that the laborer "feels that he becomes an established employee of the department and takes pride in the fact." Report on the Transport Department for the Year 1908, s. 5, PRO CO 98/16.

105. Transport Circular No. 3, Official Papers on Labour Policy, July 11, 1903, CUL RCMS 139/12/26.

106. Regulations for Registration of Labour, July 11, 1903.

107. See Colin Newbury, "Nathan, Sir Matthew (1862–1939)," *Oxford Dictionary of National Biography* (Oxford: Oxford University Press, 2004); online edition January 2008. http://www.oxforddnb.com/view/article/35189.

108. Chief Justice Griffith also rejected the bill on account of its lack of precision, which made it too far reaching. Griffith insisted that "the scheme cannot, in the circumstances in which it will be placed, be worked with that smoothness and accuracy which will be necessary to ensure success." Also, the colonial secretary wrote to Migeod on November 16, 1903, to point out that the bill would have to cover not only workers in the mines but also those employed for the railway or the Public Works Department. Otherwise, "all trace of [the worker] will be lost." If other wage laborers were not included, around Tarkwa especially, "the value of the ordinance will be completely nullified." Colonial Secretary to Migeod, November 16, 1903, RCMS 139/4/42. "The evil complained of is confined to mining labourers but the government must go further and provide for the registration of all persons employed on a concession area, from mine managers and medical

men to clerks and cooks. Of course, if the bill is made to apply to native clerks it must be extended to European employers. But there is no need to extend it to any clerks. The bill could, [I] think without difficulty, be made to apply solely to mining labourers in the same way the master and servant ordinance is made to apply solely to certain classes of servants and I will assume that the bill will affect mining labourers only." Memorandum on the Concessions Labour Bill, March 25, 1904.

109. Memorandum on the Concessions Labour Bill.
110. Memorandum on the Concessions Labour Bill.
111. Memorandum on the Concessions Labour Bill.
112. Memorandum on the Concessions Labour Bill.
113. Memorandum on the Concessions Labour Bill.
114. Memorandum on the Concessions Labour Bill.
115. Memorandum on the Concessions Labour Bill.
116. Memorandum on the Concessions Labour Bill.
117. Memorandum on the Concessions Labour Bill.
118. Memorandum on the Concessions Labour Bill.
119. Memorandum on the Concessions Labour Bill.
120. Memorandum on the Concessions Labour Bill.
121. Memorandum on the Concessions Labour Bill.
122. Memorandum on the Concessions Labour Bill.
123. Memorandum on the Concessions Labour Bill.
124. Memorandum on the Concessions Labour Bill.
125. Memorandum on the Concessions Labour Bill.
126. Memorandum on the Concessions Labour Bill.
127. Memorandum on the Concessions Labour Bill.
128. Memorandum on the Concessions Labour Bill.
129. Memorandum on the Concessions Labour Bill.
130. Memorandum on the Concessions Labour Bill.
131. Memorandum on the Concessions Labour Bill.
132. Memorandum on the Concessions Labour Bill.
133. Memorandum on the Concessions Labour Bill.
134. Memorandum on the Concessions Labour Bill.

135. This amendment made special provisions with respect to advances. However, this improvement was also built on a foundation of noncompetition among the various companies in Wassa. See subsections 39a and 39c of Ordinance 12 of 1902, Master and Servant Ordinance, 2:897.

136. Memorandum on the Concessions Labour Bill, CUL RCMS 139/12/11.

137. Memorandum on the Concessions Labour Bill, CUL RCMS 139/12/11.

138. Memorandum on the Concessions Labour Bill, CUL RCMS 139/12/11.

139. Report on the Transport Department for the Year 1904, s. 4, PRO CO 98/14.

140. Draft of Transport Circular No. 4, CUL RCMS 139/12.

141. Draft of Transport Circular No. 4.

142. For example, assistance from the agency was necessary in September 1908 when work had begun on the Tarkwa–Prestea railway. A good many requisitions from the mines came in. "Mendis [sic] were principally supplied, and nearly all went to work underground. Two gangs, numbering 50 in all, were collected by this department at Coomassie for mines in the Tarkwa district." Report on the Transport Department for the Year 1908, s. 8, PRO CO 98/16.

Chapter Five

1. Report on the Transport Department for the Year 1906, s. 5, PRO CO 98/16.

2. Thomas, "Forced Labour in British West Africa."

3. Allan McPhee, *The Economic Revolution in British West Africa* (London: Frank Cass & Co., 1926), 23, 55.

4. Lentz, *Ethnicity and the Making of History in Northern Ghana*, 139.

5. Report on the Northern Territories for the Year 1906, s. 9, PRO CO 98/16.

6. Report on the Northern Territories for the Year 1906, s. 9.

7. Thomas, "Forced Labour in British West Africa."

8. Report on the Transport Department for the Year 1906, s. 5, PRO CO 98/16; RHO MS 7ss.14 s.14.

9. Report on the Northern Territories for the Year 1906, s. 9, PRO CO 98/16.

10. Report on the Gold Mining Industry for the Year 1906, s. 5, PRO CO 98/16.

11. Thomas, "Forced Labour in British West Africa," 81.

12. Report on the Northern Territories for the Year 1906, s. 9, PRO CO 98/16.

13. PRAAD ADM 56/1/84, No. 155/7/1914; Gold Coast Departmental Report for the Year 1915, 28, PRAAD ADM 5/1/23.

14. Report on the Northern Territories for the Year 1906, s. 9, PRO CO 98/16.

15. Louis Patrick Bowler, *Gold Coast Palaver* (London: J. Long, Ltd., 1911), 84.

16. Curle, "West African Mines," 44.

17. Curle, "West African Mines," 44.

18. Thomas, "Forced Labour in British West Africa," 92.

19. Thomas, "Forced Labour in British West Africa," 79–103.
20. Thomas, "Forced Labour in British West Africa," 81.
21. Thomas, "Forced Labour in British West Africa," 86n.
22. Provincial Commissioner's Office Northern Territories, 6 March 1915, PRAAD ADM 56/1/84, No. 155/7/1914.
23. The authority of chiefs over migrant laborers was frequently simplified and exaggerated in European accounts, as demonstrated in the following report by a mine manager in Asante: "I must not omit to tell you that the native labourers before [Christmas] and sometimes as early as November leave their work for home districts, returning again at the end of January, or early in February and my own experience of this country is that nothing will induce them to remain if their Kings or their chiefs send word for them to come to pay tribute either to a king or chief lately deceased, or to even someone who has been dead many years." Report by C. W. Mann, LMA MS14171, 2:25. Contemporary observers tended to overlook familial or other social obligations related to the above mentioned events.
24. Lentz, *Ethnicity and the Making of History in Northern Ghana*, 139.
25. Lentz, *Ethnicity and the Making of History in Northern Ghana*, 139.
26. CCNT to Acting Colonial Secretary, December 2, 1909, PRAAD ADM 56/1/84.
27. Guggisberg, *We Two in West Africa*, 1:98.
28. Guggisberg, *We Two in West Africa*, 1:98. This arrangement was approved without much controversy, though there were minor worries that it appeared to have similar mechanisms to forced labor.
29. Russell, "Pidgin-English Amongst the Khroos," 206.
30. CCNT to Acting Colonial Secretary, December 2, 1909, PRAAD ADM 56/1/84.
31. The commissioner did not hide the fact that the wealth of returned laborers would "be a great help to the recruiting officer." CCNT Tamale to CCA Kumasi, August 22, 2018, PRAAD ADM 56/1/84, 8/11/19.
32. Extract from Gambaga Diary, November 12, 1917, PRAAD ADM 56/1/177.
33. CCNT to Acting Colonial Secretary, PRAAD ADM 56/1/84.
34. Extract from Gambaga Diary, November 12, 1917, PRAAD ADM 56/1/177.
35. Secretary for Mines Frank Cogill to Colonial Secretary, Secretary for Mines Office, Mines Department, November 22, 1909, No. 1096/1909.
36. Secretary for Mines Frank Cogill to Colonial Secretary, No. 1096/1909.
37. Secretary for Mines Frank Cogill to Colonial Secretary, No. 1096/1909.
38. Secretary for Mines Frank Cogill to Colonial Secretary, No. 1096/1909.

39. CCNT Tamale to Ford at Abbontiakoon Mines," August 19, 1916, PRAAD ADM 56/1/84.

40. Migrants also had their own hesitations when it came to bringing their wives to the male-dominated mining villages. Many of them complained that their wives were harassed by other men. According to one administrator, "With regard to their wives every effort is made to induce them to take them, so far only with partial success as they fear the Ashanti and Coast natives." CCNT to Acting Colonial Secretary, PRAAD ADM 56/1/84. Another European witness to these tensions reported that every "day brings a palaver amongst your Fantee labourers and the people of the village, thefts and misconduct with others' wives, which, fortunately, is always arranged by a money consideration." Bowler, *Gold Coast Palaver*, 18. A 1916 letter from the Migeod collection details the fate of one woman after separating from her migrant laborer husband. She was relegated to staying with another man in the transport department camp for porters in Kumasi. As a foreigner with no means of returning home, she begged to remain in the camp, where the girlfriends of other laborers had settled in. She wrote to Migeod: "Dear Sir, I beg most respectfully to put before you that I have got a friend called Momoh at Carrier Lines with whom I am staying, and as other carriers have friends staying with them, I would earnestly beg that you give me permission to stay. As you are aware I never left Demoe [husband] who now reported my staying in the line hut—he left me. Had the separation originated by me then one would think I want to make myself foolish, but on the contrary. Such being the case Sir, I append for your fatherly consideration. Seeing I am a woman, a stranger besides, wishing this will meet your kind approval. Yours sincerely, Abeasie (?) her x mark." Abeasie (?) to Migeod, CUL RCMS 139/16.

41. Lentz, "Colonial Constructions and African Initiatives: The History of Ethnicity in Northwestern Ghana," *Ethnos* 65 (2001): 127.

42. Thomas, "Forced Labour in British West Africa." See also Roger Thomas, "Military Recruitment in the Gold Coast during the First World War," *Cahiers d'Études africaines* 15 (1975): 62.

43. Gold Coast Departmental Report for the Year 1916, s. 4, PRAAD ADM 5/1/24.

44. Report on the Northern Territories for the Year 1906, s. 9, PRO CO 98/16.

45. Lentz, *Ethnicity and the Making of History in Northern Ghana*, 140.

46. For example, "Although in July 1909 a total of 444 migrants arrived In Tarkwa, only 363 migrants returned to Wa a year later, 16 having remained in Tarkwa, 22 having 'deserted,' almost 30 not yet having reported to the district commissioner and 15 having died." Lentz, *Ethnicity and the Making of History in Northern Ghana*, 141. Prior to this wave of recruitment, mine managers had attempted to bar employers on the railway from recruiting

around the mines and to get the government to enforce wage rates for railway work which were below the mines' standards due to fears of poaching. Thomas, "Forced Labour in British West Africa," 82.

47. CCNT to Secretary for Mines," December 30, 1905, PRAAD ADM 56/1/3.

48. Report on the Northern Territories, Immigration and Emigration, Gold Coast Departmental Report for the Year 1915, 9, PRAAD ADM 5/1/23.

49. Edwin Cade, "Report to the Directors of the Cote d'Or Company," December, 10 1897, s. 9, Cade Papers.

50. Gold Coast Departmental Report for the Year 1915, 8, PRAAD ADM 5/1/23.

51. Précis of Informal Diary of CCT from December 15 to February 14, 1915, PRAAD ADM 56/1/84.

52. Gold Coast Departmental Report for the Year 1915, 8, PRAAD ADM 5/1/23.

53. For instance, in 1909 Secretary of Mines Cogill and Lieutenant Colonel Watherson shared a proposed "bill on native labour" with the attorney general, proposing a long list or regulations: concerning workers' housing; the method of pay; punishment and fines (also for deserters), market prices for food stuffs; duration of contract and under which circumstances labor is considered to be employed; forcing employers to recruit through the newly constructed labor bureau; the establishment of a board of control led by the secretary of mines, directors of the labor bureau, and the governor, giving the governor powers to make changes to the master and servant ordinance. Secretary of Mines and Chief Commissioner to Honourable Attorney General, August 28, 1909, ADM 56/1/84.

54. A handwritten version of the "regulations and instruction as to the supply of labourers from the Northern Territories of the Gold Coast to the mines" is located at the National Archives in Accra, Ghana. PRAAD ADM 56/1/84. See also Draft Rules for Mine Labour by CCNT, ADM 56/1/84; RCMS 139/12/14.

55. CCNT to Colonial Secretary, February 15 1910, PRAAD ADM 56/1/84 No. 7/1208/1910.

56. Draft Rules for Mine Labour by CCNT, ADM 56/1/84.

57. Draft Rules for Mine Labour by CCNT, RCMS 139/12/14.

58. Draft Rules for Mine Labour by CCNT, RCMS 139/12/14.

59. In May of 1909, Giles Hunt, an attorney for some of the biggest mines in Wassa, sent yet another proposal to the administration for a labor bureau with a pass law system, this time in conjunction with a compound structure under European supervision as it had been introduced on the Witwatersrand. PRAAD ADM 56/1/84.

Conclusion

1. Ulbe Bosma, Elise van Nederveen Meererk, and Aditya Sarkar, "Mediating Labour: An Introduction," *International Review of Social History* 57 (2012): 1–15.

Bibliography

Primary Sources

Archival Sources

Archives and Special Collections of the School of African and Oriental Studies, London.
British Parliamentary Papers (PP).
Cade Papers, Centre of West African Studies, University of Birmingham, United Kingdom.
Digitized Books from the University of Illinois at Urbana-Champaign and the Open Content Alliance, United States.
Indiana University Archives, Bloomington, United States.
London Metropolitan Archives, Guildhall, London (LMA).
National Archives, London (PRO).
Papers of Frederick William Hugh Migeod, Royal Commonwealth Society, Cambridge University Library (CUL).
Public Records and Archives Administration Department of Ghana (PRAAD) in Accra and the Regional Archives, Sekondi, Ghana.
Rhodes House Library, University of Oxford (RHO).

Published Primary Sources

"Abbontiakoon (Wassaw) Mines Limited," *Economist*, May 4, 1901.
"Abbontiakoon (Wassaw) Mines Limited," *Economist*, November 15, 1902.
Allen, William, and T. R. H. Thomson. *A Narrative of the Expedition Sent by Her Majesty's Government to the River Niger in 1841: Under the Command of Captain H. D. Trotter, R.N.* 2 vols. London: Frank Cass & Co Ltd., 1841.
Bonnat, Marie-Joseph. *Marie-Joseph Bonnat et les Ashanti, 1869–1874.* Paris: Société des Africanistes, 1994.
Bowler, Louis Patrick. *Gold Coast Palaver.* London: J. Long, Ltd., 1911.
Burton, Richard F. Letter to the editor, *Mining World and Engineering Record*, March 17, 1883, p. 283.
Burton, Richard F., and Verney L. Cameron. *To the Gold Coast for Gold: A Personal Narrative.* 2 vols. London, 1883.

Buell, Raymond Leslie. *The Native Problem in Africa.* 2 vols. New York: Macmillan, 1926.

"Company Report of the Taquah Mining & Exploration," *Mining Magazine* 2 (1910).

"Company Report of the Taquah Mining & Exploration," *Mining Magazine* 5 (1911).

Curle, J. H. "West African Mines." *Mining Magazine* 1 (1909): 42–46.

"Fanti Consolidated Mines, Limited." *Economist,* February 16, 1901, 249.

Fitzgerald, Ferdinand. "The Gold Coast Colony Lands." *African Times,* April 1, 1875, 43.

———. "The Gold Coast Colony Roads and Seaport." *African Times,* June 1, 1875, 69.

"Gold Coast Investment Company Limited." *Economist,* March 23, 1901.

"The Gold Coast, West Africa." *Kalgoorlie Miner,* September 17, 1901, p. 3.

Great Britain Colonial Office. *Colonial Office List for 1920.* London: Colonial Office, 1920.

Guggisberg, Decima Moore. *We Two in West Africa.* 2 vols. New York: W. Heinemann, 1909.

Horton, James Africanus Beale. "The West Africa Gold Fields." *African Times,* December 1, 1877, 139.

"The 'Lloyd' Copper Company, Ltd." *Financial Times,* May 29, 1899.

Loring, W. J. "The West African Outlook." *Kalgoorlie Western Argus,* August 3, 1909, p. 8.

McCarthy, E. T. "Early Days on the Gold Coast." *Mining Magazine* 1 (December 1909): 291–94.

McPhee, Allan. *The Economic Revolution in British West Africa.* London: Frank Cass & Co., 1926.

Milne, Alan Hay. *Sir Alfred Lewis Jones K. C. M. G.: A Story of Energy and Success.* Liverpool: Henry Young & Sons, 1914.

"Only a Kru-boy, But . . ." *Work and Workers in the Mission Field.* Wesleyan Methodist Missionary Society, 1920.

Ramseyer, Friedrich August, and Johannes Kühne. *Four Years in Ashantee.* Summit, NJ: J. Nisbet & Co., 1878.

Russell, J. D. "Pidgin-English amongst the Khroos." *Work and Workers in the Mission Field.* Wesleyan Methodist Missionary Society, 1909.

Starr, Frederick. *Liberia: Description, History, Problems.* Chicago: N.p., 1913.

Stockfeld, Gerhard. "Abosso Gold." *Financial Times,* December 16, 1905.

"Taquah and Abosso Gold Mining Company (1900) Limited." *Economist,* December 16, 1905.

Thornhill, John Bensley. *Adventures in Africa under the British, Belgian and Portuguese Flags.* London: John Murray, 1915.

"Wassau (Gold Coast) Mining Company, Limited." *Economist,* March 23, 1901, 452.

Way, Herbert W. L. *Round the World for Gold: A Search for Minerals from Kansas to Cathay.* London: Sampson Low, Marston & Company, Ltd., 1912.
"West African Finance." *Economist,* December 24, 1904: 2095.
"West African Mining." *Mining Magazine* (July–December 1910): 3:322.
"West African Results." *Economist,* January 3, 1903.
Wills, Walter H., and Barrett, R. J. *The Anglo-African Who's Who and Biographical Sketchbook.* London: George Routledge & Sons, Ltd., 1905.

Secondary Sources

Adzoxorno, Isaacus. "The Origin and Development of the Individual Contract of Employment in Ghanaian Labour Law." Paper delivered at the 1987 Conference of the African Studies Association of Australasia and the Pacific, Adelaide University, August 1987.
Ahuja, Ravi. "Mobility and Containment: Voyages of South Asian Seamen, c. 1900 to 1960." In *Coolies, Capital and Colonialism: Studies in Indian Labour History,* edited by Rana Behal and Marcel van der Linden, 111–42. Cambridge: Cambridge University Press, 2007.
Akurang-Parry, Kwabena O. "African Agency and Cultural Initiatives in the British Imperial Military and Labor Recruitment Drives in the Gold Coast (Colonial Ghana) during the First World War." *African Identities* 4 (2006): 213–34.
———. "Colonial Forced Labor Policies for Road-Building in Southern Ghana and International Anti-forced Labor Pressures, 1900–1940." *African Economic History* 28 (2000): 1–25.
———. "'The Loads Are Heavier Than Usual': Forced Labor by Women and Children in the Central Province, Gold Coast (Colonial Ghana), c. 1900–1940." *African Economic History* 30 (2002): 31–51.
———. "'We Cast about for a Remedy': Chinese Labor and African Opposition in the Gold Coast, 1874–1914." *International Journal of African Historical Studies* 34 (2001): 365–84.
Akurang-Parry, Kwabena Opare. "Missy Queen in Her Palaver Says De Gole Cosse Slaves Is Free: The British Abolition of Slavery/Pawnship and Colonial Labor Recruitment in the Gold Coast [Southern Ghana], 1874–Ca. 1940." PhD diss., York University, 1999.
Alexander, Peter. "Challenging Cheap-Labour Theory: Natal and Transvaal Coal Miners, ca. 1890–1950." *Labor History* 49 (2008): 47–70.
Allen, G. Keith. "Gold Mining in Ghana." *African Affairs* 57 (1958): 221–40.
Allman, Jean. "Let Your Fashion Be in Line with Our Ghanaian Costume: Nation, Gender, and the Politics of Clothing in Nkrumah's Ghana." In *Fashioning Africa: Power and the Politics of Dress,* 144–65. Bloomington: Indiana University Press, 2004.

Austin, Gareth. "Cash Crops and Freedom: Export Agriculture and the Decline of Slavery in Colonial West Africa." *International Review of Social History* 54 (2009): 1–37.

———. *Labor, Land, and Capital in Ghana: From Slavery to Free Labor in Asante, 1807–1956*. Rochester, NY: University of Rochester Press, 2005.

———. "Vent for Surplus or Productivity Breakthrough? The Take-off of Ghanaian Cocoa Exports, c. 1890–1936." *Economic History Review* 64 (2014): 1035–64.

Barchiesi, Franco. "How Far from Africa's Shore? A Response to Marcel van Der Linden's Map for Global Labor History." *International Labor and Working-Class History* 82, no. 3 (2012): 77–84.

Barchiesi, Franco, and Stefano Bellucci. "Introduction." *International Labor and Working-Class History* 86 (2014): 4–14.

Behrens, Christine. *Le Croumen de la côte occidental de l'Afrique*. Bordeaux: Ministère de l'Éducation Nationale, 1974.

Bosma, Ulbe, Elise van Nederveen Meererk, and Aditya Sarkar. "Mediating Labour: An Introduction." *International Review of Social History* 57 (2012): 1–15.

Brass, Tom, and Marcel van der Linden, eds. *Free and Unfree Labour: The Debate Continues*. Bern: Peter Lang AG, 1997.

Brown, Carolyn A. *"We Were All Slaves": African Miners, Culture, and Resistance at the Enugu Government Colliery*. Cape Town: James Currey, 2003.

Brown, Carolyn A., and Marcel van der Linden. "Shifting Boundaries between Free and Unfree Labor: Introduction." *International Labor and Working-Class History* 78, no. 1 (2010): 4–11.

Burbank, Jane, and Frederick Cooper. "Imperial Trajectories." In *Empires in World History: Power and the Politics of Difference*, 1–23. Princeton, NJ: Princeton University Press, 2010.

Chada, G. T. Z. "Labor Protests, Group Consciousness and Trade Unionism in West Africa: The Radical Railway Workers of Colonial Ghana." PhD diss., University of Toronto, 1981.

Clarence-Smith, Gervase. "Thou Shalt Not Articulate Modes of Production." *Canadian Journal of African Studies / Revue Canadienne des Études Africaines* 19 (1985): 19–22.

Cooper, Frederick. "African Labor History." In *Global Labor History: A State of the Art*, edited by Jan Lucassen, 91–116. Bern: Peter Lang AG, 2006.

———. *Colonialism in Question: Theory, Knowledge, History*. Berkeley: University of California Press, 2005.

Crisp, Jeff. "The Labor Question on the Gold Coast." In *Proletarianization in the Third World: Studies in the Creation of a Labor Force under Dependent Capitalism*, edited by Henry Finch and Barry Munslow, 18–42. London: Routledge, 1984.

———. "Productivity and Protest: Scientific Management in the Ghanaian Gold Mines, 1947–1956." In *Struggle for the City: Migrant Labor, Capital, and the State in Urban Africa*, edited by Frederick Cooper, 91–130. Beverly Hills, CA: SAGE Publications, 1983.

———. *The Story of an African Working Class: Ghanaian Miners' Struggles, 1870–1980*. London: Zed Books, 1984.

Dickson, Kwamina B. *A Historical Geography of Ghana*. Cambridge: Cambridge University Press, 1969.

Dorward, David. "'Nigger Driver Brothers': Australian Colonial Racism in the Early Gold Coast Mining Industry." *Ghana Studies*, 5 (2002): 197–214.

Dumett, Raymond. *El Dorado in West Africa: The Gold-Mining Frontier, African Labor, and Colonial Capitalism in the Gold Coast, 1875–1900*. Athens, OH: Ohio University Press, 1998.

———. "The Nzemans of Southwestern Ghana: Gold Miners, Rubber Traders, Loggers and Entrepreneurs." In *Ghana in Africa and the World: Essays in Honor of Adu Boahen*, edited by Toyin Falola, 455–76. Trenton, NJ: Africa World Press, 2003.

———. "Obstacles to Government-Assisted Agricultural Development in West Africa: Cotton-Growing Experimentation in Ghana in the Early Twentieth Century." *Agricultural History Review* 23 (1975): 156–72.

———. "Parallel Mining Frontiers in the Gold Coast and Asante in the Late 19th and Early 20th Centuries." In *Mining Frontiers in Africa: Anthropological and Historical Perspectives*, edited by Katja Werthmann and Tilo Grätz, 33–54. Köln: Rüdiger Köppe Verlag, 2012.

Dumett, Raymond, and Marion Johnson. "Britain and the Suppression of Slavery in the Gold Coast Colony, Ashanti, and the Northern Territories." In *The End of Slavery in Africa*, edited by Suzanne Miers and Richard Roberts, 71–116. Madison: University of Wisconsin Press, 1988.

Eckert, Andreas. "What is Global Labour History Good For?" In *Work in a Modern Society: The German Historical Experience in Comparative Perspective*, edited by J. Kocka, 169–82. New York: Berghahn Books, 2013.

Engelken, Dagmar. "A White Man's Country: The Chinese Labor Controversy in the Transvaal." In *Wages of Whiteness and Racist Symbolic Capital*, edited by Wulf D. Hund, Jeremy Krikler, and David R. Roediger, 161–94. Berlin: LIT Verlag, 2010.

Fall, Babacar. *Social History in French West Africa: Forced Labour, Labour Market, Women and Politics*. Amsterdam: Sephis, 2002.

Ferguson, James. *Global Shadows: Africa in the Neoliberal World Order*. Durham, NC: Duke University Press, 2006.

Findlay, George G. *The History of the Wesleyan Methodist Missionary Society*. 5 vols. London: Epworth Press, 1921.

Frankema, Ewout, and Marlous van Waijenburg. "Structural Impediments to African Growth? New Evidence from Real Wages in British Africa, 1880–1965." *Journal of Economic History*, 72, no. 4 (December 2012): 895–926.

Freund, Bill. *Capital and Labour in the Nigerian Tin Mines*. Atlantic Highlands, NJ: Longman, 1981.

Frost, Diane. *Work and Community among West African Migrant Workers since the Nineteenth Century*. Liverpool: Liverpool University Press, 1999.

Gann, L. H., and Peter Duignan. *The Rulers of British Africa, 1870–1914*. Stanford: Stanford University Press, 1978.

Gaunt, Mary. *Alone in West Africa*. New York: Charles Scribner's Sons, 1911.

Geschiere, Peter. "Chiefs and Colonial Rule in Cameroon: Inventing Chieftaincy, French and British Style." *Africa: Journal of the International African Institute* 63 (1993): 151–75.

"Gold: Akan Goldfields: 1400 to 1800." In *Encyclopedia of African History*, edited by Kevin Shillington, 1:586–87. New York: Taylor & Francis Group, 2004.

Hamill, John. *The Strange Career of Mr. Hoover under Two Flags*. New York: J. Day Co., 1931.

Hill, Polly. *Development Economics on Trial: The Anthropological Case for a Prosecution*. Cambridge: Cambridge University Press, 1986.

———. *The Migrant Cocoa-Farmers of Southern Ghana: A Study in Rural Capitalism*. Cambridge: Cambridge University Press, 1997.

Hofmeester, Karin, Jan Lucassen, and Filipa Ribeiro da Silva. "No Global Labor History without Africa: Reciprocal Comparison and Beyond." *History in Africa* 41 (2014): 249–76.

Jeeves, Alan. *Migrant Labor in South Africa's Mining Economy: The Struggle for the Gold Mines' Labor Supply, 1890–1920*. Kingston, ON: McGill-Queen's University Press, 1985.

Jenkins, Paul. "The Anglican Church in Ghana, 1905–24 (I)." *Transactions of the Historical Society of Ghana* 1 (June 1974): 25–26.

Keese, Alexander. "Forced Labour in the 'Gorgulho Years': Understanding Reform and Repression in Rural São Tomé e Príncipe, 1945–1953." *Itinerario* 38, no. 1 (2014): 103–24.

———. "Slow Abolition within the Colonial Mind: British and French Debates about 'Vagrancy,' 'African Laziness,' and Forced Labour in West Central and South Central Africa, 1945–1965." *International Review of Social History* 59, no. 3 (2014): 377–407.

Killingray, David. "Labour Exploitation for Military Campaigns in British Colonial Africa 1870–1945." *Journal of Contemporary History* 24 (July 1989): 483–501.

———. "Labor Mobilization in British Colonial Africa for the War Effort, 1936–46." In *Africa and the Second World War*, edited by David Killingray and Richard Rathbone, 68–96. London: Palgrave Macmillan, 1986.

Kimble, David. *A Political History of Ghana: The Rise of Gold Coast Nationalism, 1850–1928*. Oxford: Oxford University Press, 1963.

Lawrance, Ben N., Emily Lynn Osborn, and Richard L. Roberts. "Introduction: African Intermediaries and the 'Bargain' of Collaboration." In *Intermediaries, Interpreters, and Clerks: African Employees in the Making of Colonial Africa*. Madison: University of Wisconsin Press, 2006.

Lentz, Carola. "The Chief, the Mine Captain and the Politician: Legitimating Power in Northern Ghana." *Africa: Journal of the International African Institute* 68 (1998): 46–67.

———. "Colonial Constructions and African Initiatives: The History of Ethnicity in Northwestern Ghana." *Ethnos* 65 (2001): 107–36.

———. *Ethnicity and the Making of History in Northern Ghana*. Edinburgh: Edinburgh University Press, 2006.

———. *Land, Mobility, and Belonging in West Africa*. Bloomington: Indiana University Press, 2013.

Lentz, Carola, and Veit Erlmann. "A Working Class in Formation? Economic Crisis and Strategies of Survival among Dagara Mine Workers in Ghana." *Cahiers d'Études africaines* 29 (1989): 69–111.

Maier, Gerhard. *African Dinosaurs Unearthed: The Tendaguru Expeditions (Life of the Past)*. Bloomington: Indiana University Press, 2003.

Martin, Jane. "Krumen 'Down the Coast': Liberian Migrants on the West African Coast in the 19th and early 20th Centuries." *International Journal of African Historical Studies*, 18, no. 3 (1985): 401–23.

Martino, Enrique. "Clandestine Recruitment Networks in the Bight of Biafra: Fernando Pó's Answer to the Labour Question, 1926–1945." *International Review of Social History* 57 (2012): 39–72.

Metcalfe, George Edgar. *Great Britain and Ghana: Documents of Ghana History 1807–1957*. London: Thomas Nelson, 1964.

Newbury, Colin. "Nathan, Sir Matthew (1862–1939)," *Oxford Dictionary of National Biography*. Oxford: Oxford University Press, 2004, http://www.oxforddnb.com/view/article/35189.

Phillips, Anne. *The Enigma of Colonialism: British Policy in West Africa*. London: James Currey, 1989.

Phimister, Ian, and Jeremy Mouat. "Mining, Engineers and Risk Management: British Overseas Investment, 1894–1914." *South African Historical Journal* 49 (2003): 1–26.

Rosenblum, Paul. "Gold Mining in Ghana 1874–1900." PhD diss., Columbia University, 1972.

Shumway, Rebecca. *Fante and the Transatlantic Slave Trade*. Rochester, NY: University of Rochester Press, 2011.

Simpson, George Gaylord. *Simple Curiosity: Letters from George Gaylord Simpson to His Family, 1921–1970*. Berkeley: University of California Press, 1987.

Sundiata, Ibrahim. *From Slaving to Neoslavery: The Bight of Biafra and Fernando Po in the Era of Abolition, 1827–1930*. Madison: University of Wisconsin Press, 1996.

Swindell, Kenneth, and Alieu Jeng. *Migrants, Credit and Climate: The Gambian Groundnut Trade, 1834–1934*. Leiden: Brill, 2006.

Szereszewski, Robert. *Structural Changes in the Economy of Ghana, 1891–1911*. London: Weidenfeld & Nicolson, 1965.

Thomas, Roger. "Forced Labour in British West Africa: The Case of the Northern Territories of the Gold Coast 1906–1927." *Journal of African History* 14 (1973): 79–103.

———. "Military Recruitment in the Gold Coast during the First World War." *Cahiers d'Études africaines* 15 (1975): 57–83.

Trapido, Stanley. "South Africa in a Comparative Study of Industrialization." *Journal of Development Studies* 7, no. 3 (1971): 309–20.

van der Linden, Marcel. "The Origins, Spread and Normalization of Free Wage Labour." In *Free and Unfree Labour: The Debate Continues*, edited by Tom Brass and Marcel van der Linden, 501–23. Bern: Peter Lang AG, 1997.

———. "Plädoyer für eine Historische Neubestimmung der Welt-Arbeiterklasse." *Sozial Geschichte: Zeitschrift für historische Analyse des 20. und 21. Jahrhunderts* 20 (2005): 7–28.

———. "Warum Gab (und Gibt) es Sklaverei im Kapitalismus? Eine einfache und dennoch schwer zu beantwortende Frage." In *Unfreie Arbeit: Ökonomische und Kulturgeschichtliche Perspektiven*, edited by M. Erdem Kabadayi and Tobias Reichardt, 260–79. Hildesheim: Georg Olms, 2007.

Weber, Max. *General Economic History*. Translated by Frank H. Knight. New York: Greenberg Publishers, 1927. Reprint, New York: Dover, 2012.

Williamson, Thora. *Gold Coast Diaries: Chronicles of Political Officers in West Africa, 1900–1919*. Edited by Anthony Kirk-Greene. London: Radcliffe Press, 2000.

Index

Page numbers in italics signify graphics or tables.

Abbontiakoon Mines, 20, 23, 89, 148–49, 155; production at, 44–46
Aborigines Protection Society, 123, 139
Abosso, 22, 48, 130
Abosso Gold Mining Company, 20, 47, 59, 124; and instrumentalization of debt, 90–92; and labor recruitment, 50, 109, 126; production at, 48–50
Accra, Coffi, 59
Adjah Bippo, 37, 42, 57, 58, 73
advance payment system, 85, 90, 143, 151–52, 159; and labor regulation, 86–93, 127–28. *See also* cash advances; extended payment system
African Estates Agency, 41
African Gold Coast Mining Company, 19, 20, 48, 58
African Review, 42
African Times, 23, 26, 28
Aggrey, John, 132–34
agrarian credit, 169n37
Akan people: as labor agents, 109; as laborers, 8, 22, 52, 54–55, 59; as moneylenders, 162
Akurang-Parry, Kwabena, 64, 177n47, 182n131
Alimendi (contract laborer), 97–98

Allen, William, 68–69, 75–76, 99
Allman, Jean, 77
Amman, Quamina, 109–10
Anfargah Mining Company, 133–34
Apinto, 29, 30, 42
Apollonians (Nzemans), 55, 56, 63, 107
Asante kingdom, 19–20, 34; British war against, 24, 29, 30
Asante laborers, 55, 107
Asante region, 131
Ashanti Gold Fields Corporation, 12, 102, 112, 113–14
Asian laborers: and Africans, 63, 64, 177n47; illness and disease suffered by, 63–64; mine employers' desire for, 54, 61–64, 120–22
Austin, Gareth, 166n14, 167n20, 181n121
Australia, 34–35, 121

backward bending supply curve, 7, 167n20
Bailey, George, 48
Benusan, E. V., 127
Bomah, Joseph, 98–99
Bonnat, Marie-Joseph, 19–20
Bowden, Arthur, 59
Britain: colonial rule by, 19, 24–26, 146–47, 150–51; government of, 32–33. *See also* Colonial Office

British Gold Coast Company, 42
Brown, Carolyn, 67
Bryan, Herbert, 86
Buchanan, Thomas, 68–69
Buell, Raymond, 103
Bullen, James, 106, 123, 184n13
Burbank, Jane, 9
Burton, Richard Francis, 21, 22–23, 56–57, 62–63, 178n59
Bwana Mkubwa, 41

Cade, Edwin, 102
Cameron, Verney Lovett, 21, 56–57, 62–63, 178n59
Cape Coast, 110, 125
capitalism, 4, 8, 9, 26, 162; and labor, 10–11, 168n32
cash advances, 65, 69, 71, 77, 90, 131; and debt bondage, 83, 85–86. *See also* advance payment system; extended payment system
casual labor, 54–60, 183n24
Chada, G. T. Z., 181n131
Chamberlain, Joseph, 33
cheap labor theory, 8, 168n25, 178n59
Cheeseman, Joseph James, 103
chiefs: authority of, 6, 151, 196n23; as labor agents, 145–47; and labor recruitment, 147–48, 149–51, 155; and tributary labor, 56
Chinese labor, 54, 62–63, 121, 187n7. *See also* Asian laborers
Cinnamon Bippo, 37, 42, 126; about, 21–22; advance payments at, 90–91; and tributary labor, 57, 58
Clarence-Smith, Gervase, 11
class formation, 4–5, 11, 93, 158, 163
coal, 46

coercion, 6, 59, 77, 150, 161; and capitalist labor relations, 11, 168n32; and contract labor from Liberia, 64–68; and extended payment system, 90–91, 159–60. *See also* forced labor
coffee prices, 106, 184n11
Cogill, Frank, 145, 148, 152–53, 155, 156–57, 198n53
Colonial Office, 22, 24, 44; and government transport office, 124, 125; and labor discipline, 87; and labor recruitment, 5–6, 106, 121–23; and mining interests, 27, 32, 40, 106, 122–23, 125, 138; staff changes in, 32–33
colonialism, 135; British, 19, 24–26, 146–47, 150–51; Dutch, 1, 24; French, 101–3; Portuguese, 3, 107
Concession Ordinance of 1900, 125, 189n28
Concessions Labor Ordinance of 1903, 137–44, 193–94n108
Consolidated Gold Fields, 41, 46, 121, 123
contested collaboration, 9
contract labor, 11, 52–53, 65, 87, 93, 114–15. *See also* laborers, African
Cooper, Frederick, 9
corporal abuse, 133–34
Creswell, Frederick Hugh Page, 120–21, 187n8
Crisp, Jeff, 68
Crocker, Frederick J., 22, 57, 58
Crompton, W. L., 72
Curle, James Herbert, 149
Cuscaden, William Andrew "Tim," 27–31, 58

Dagarti, Isaka, 117–18
Dahse, Paulus, 23, 30–31, 34

Dalglish, Thomas F., 48
Davis, Edmund, 40–42
Davis, Tarbutt, Janson Group, 44
Dawson, Joseph, 23, 30
debt bondage, 11, 85–86, 90–93, 159–60
deep-level mining: development of, 40, 48–49, 119; and labor force, 49, 108; Tarbutt as proponent of, 41
desertion, 90–91, 154–55, 197n46; causes of, 88–89, 116, 132; and government transport office, 127–28; Griffith proposals on, 142–44
Dickinson, J. R., 68
Dorward, David, 173n83
Dumett, Raymond, 3–4, 32; on indigenous population, 2, 167n18; on labor force, 7–8, 11, 168n25
Dunkwa, 118, 125, 189n31
Dutch, 1, 24

Eamonson, L., 112–14, 162
Economist, 37, 42
Effuenta Gold Mining Company, 20, 28, 37, 39, 42, 127
El Dorado in West Africa (Dumett), 4
Emancipation Proclamation of 1874, 6, 24, 52, 87
Engelken, Dagmar, 187n7
English language, 89
ethnic identity, 4–5, 66–67, 79
European laborers. *See* white miners
Ewe people, 56, 107
extended payment system, 52, 66, 89, 181n121; as coercive tool, 90–91, 159–60; explained, 85–86, 92–93. *See also* advance payment system; cash advances

family labor, 59
family migration, 152–53, 197n40
Fante Confederation, 24, 170n25
Fanti Consolidated Mines, 73
Financial Times, 49
firewood, 45–46, 111
Firminger, Reginald E., 34
First Chimurenga, 46
Fitzgerald, Ferdinand, 23–26, 28, 30–31, 44
food, miners', 128–29, 133, 156, 191n72
forced labor, 146, 150; failure of, 154–56. *See also* coercion
Frankema, Ewout, 83, 167n22
French colonialism, 101–3
Frost, Diane, 180n113, 184n8
Fry, John, 115

gang labor system, 76–77, 79–81
gang leaders, 77, 99, 114, 115, 137; and advance payment system, 89–90, 91–92; conflicts between laborers and, 115–16, 129–30; employers' disputes with, 132–33, 134; and poaching, 117–18; power and authority of, 82, 119, 128, 129, 161–62; responsibilities of, 75, 79–81, 180n113; and wages, 129, *130*; and women, 110–11. *See also* supervisors, indigenous
Garawe, 102
Gaunt, Mary, 1, 83
Gbainya, Momo, 97
Gibson, Garretson Warner, 103
global labor history, 10–12, 14, 162, 168n32, 168-9n35, 169n36
global labor migration, 160–61, 162–63
gold boom, first (1879–85), 16, 18, 19–21; and labor question, 8, 21–26; and mechanized mining, 26–32

gold boom, second (1900–1905), 16, 18; about, 32–40; growth in number of companies during, 35, 37
Gold Coast Agency, 123
Gold Coast Colony, 102, 139, 149; abolition of slavery in, 4, 5–6, 52, 87, 166n14; Concessions Labor Ordinance in, 137–44, 193–94n108; and French colonialism, 101–3; labor laws in, 87–93; and labor mobilization, 5–6, 165–66n11; map of, *17*; Master and Servant Ordinance in, 65, 66, 70, 87–88, 92, 114, 137, 143, 161, 163, 181–82n131, 194n135; protectorates of, 19, 25, 37, 56, 111, 123. *See also* government transport department
gold mining: boom-and-bust nature of, 12; capital investments in, 26–27, 33, 41, 145; deep-level, 40, 41, 48–49, 108, 119; and drilling, 44; fuel source in, 45–46; labor shortage in, 53, 112, 120, 145; list of companies, *51*; mechanized, 26–32; panning in, 111; production output in, 3–4, *36*, 49, 172n74; production process of, 45; and railways, 1, 2, 16, 33–34, 37, 131; and Sekondi port, 1; speculative investments in, 20–21, 28; statistics on number of workers in, 2, 60, *61*, 73, *74*, 108; surface, 55–56, 58, 83, 107–8, 112, 121; traditional African, 31, 42–43, 111; transformation of labor regime in, 4–5; underground, 133, 150–51, 159
Government Gazette, 127–28, 133, 139
government transport department, 118, 119, 125–30; founding of, 125; goals of, 120; and labor recruitment, 120, 125–30, 144, 195n142; and labor registration, 137, 138–39, 141, 193n104; loan security measures of, 129–30; as mediator, 124, 130–35, 191n65; and workers' welfare, 128–29. *See also* Migeod, F. W. H.
Griffith, W. Brandford, 27–28, 33, 54, 57, 139
Griffith, William B., 139–43, 193–94n108
Guggisberg, Decima Moore, 44, 70–71, 111
Guggisberg, Frederick Gordon, 44–45

Hammah, Quao, 109–10
Higgins, Henry, 28
Hill, Polly, 11, 169n37, 183n168
hiring process, 23, 57, 65, 69, 120; and cash advances, 66, 91; and government transport office, 125, 126, 127, 128, 129, 132
Horton, James Africanus Beale, 23, 26, 30–31, 44
Humplmayr, August, 65
Hunt, Giles, 198n59
hybrid Marxism, 11

illness and disease, 63–64, 71–73, 154, 179–80n88
indentured labor, 11; from Asia, 54, 61–64, 120–22, 177n47; from Liberia, 64–68
independent labor contractors, 15, 108–14, 118; European miners as, 112–14; swindling of workers by, 114–15; women as, 110–12
indigenous labor agents: and advance payment system, 89–90, 91–92, 151–52; career-mindedness of, 69–70, 81, 99, 162; chiefs and village headmen

as, 145–47; Cuscaden on, 29–31; deception by, 66, 114–15; emergence of, 53, 68–69, 75; and ethnic identity, 79; former mine workers as, 109–10; as gang leaders, 79–81, 89, 117–18, 119, 128, 129, 161–62, 180n113; laborers' relations with, 10, 76–77, 82, 115–16, 159; Liberian contractors, 53, 65–67; mediating role of, 3, 9, 75, 79–80; and mining entrepreneurs, 68, 70, 79–80, 108, 115–16; photos of, *78*; poaching by, 116–18; power and authority of, 82, 119, 128, 129, 161–62; profile of, 70; studies on, 9–10; as supervisors, 68–75, 81, 89–90, 153, 159; women as, 101, 110–12; and workers' welfare, 79–80, 108. *See also* gang leaders; independent labor contractors

indirect recruitment, 3, 9, 100, 146, 163; indigenous control of, 15, 67, 77, 146; toleration of, 108, 114; transformations of, 118, 153, 157. *See also* labor recruitment

Intermediaries, Interpreters, and Clerks: African Employees in the Making of Colonial Africa (Lawrence, Osborn, and Roberts), 9

Irvine, James, 21–23, 28

Ivory Coast, 102, 184n4

Janson, Edmund William, 40, 41–42

Jeng, Alieu, 5

Johnson, Hilary R., 102

Jones, Alfred Lewis, 122, 188n16

"jungle boom." *See* gold boom, second

Kakari, Kofi, 20

Keese, Alexander, 165–66n11

Kofi, Albert, 117

Kru Town, 76, 106

Kuma, Enemil Kwao, 59

Kumase, 1, 33, 97, 98, 112, 125, 151

Kwamin, Bonsa, 117

labor bureau, 122, 188n15, 198n53

labor recruitment: of Asian indentured, 54, 61–64, 120–22; bottlenecks in, 101–8; brutal tactics in, 155–56; and chiefs, 147–48, 149–51, 155; coercion in, 6, 64–68, 90–91, 159–60; and colonial authorities, 5–6, 106, 121–22, 165–66n11; deceptive practices in, 66, 114–15; economic incentives for, 82–83; employer cooperation around, 123–24; and government transport office, 120, 125–30, 144, 195n142; indirect methods of, 3, 9, 15, 67, 77, 100, 108, 114, 118, 146, 153, 157, 163; labor law regulation of, 86–87; Liberia as source of, 103, 105–6, 184n8, 184n10; mainstream nature of, 75–76; and mining entrepreneurs' requests, 122–23, 148–49; new West African networks of, 101, 106–7; from Northern Territories, 145–53; and poaching, 116–18; political method of, 146, 154–55; reverting to nonpolitical method of, 156–57, 198n53; for underground mining, 150–51; voluntary nature of, 99, 140, 146; and wage labor, 76, 81; by white miners, 112–14; and workers' families, 152–53, 197n40. *See also* independent labor contractors; indigenous labor agents labor registration, 124, 144; and Concessions Labor Ordinance, 139–43, 193–94n108; government transport office initiation of, 137, 138–39, 141, 193n104

labor shortage, 53, 102, 120, 145, 161

laborers, African: and Asian laborers, 63, 64, 177n47; bargaining position of, 71, 77, 83, 113, 158; calls for registration of, 124, 137, 138–43, 141, 193n104, 193–94n108; clothing of, 77; Concessions Labor Ordinance on, 137–44; conflicts with employers by, 132–34; corporal abuse of, 133–34; desertion by, 88–89, 90–91, 116, 132, 142–44, 154–55; and disease, 63–64, 154; and ethnic identity, 4–5, 66–67, 79; and food, 128–29, 133, 156, 191n72; gang leaders' conflicts with, 115–16, 129–30; high status of, 75–76; hiring of, 23, 57, 65, 66, 69, 91, 120, 125–29, 132; indebtedness of, 11, 85–86, 90–93, 129–30, 159–60; labor agents' relations with, 10, 76–77, 82, 113, 115–16, 159, 161–62; and labor courts, 92; labor law protection of, 79; mine managers' relations with, 25, 70–71, 162; motivations of, 7, 93–100; names given to, 64–65, 177–78n53; and piecework, 58–60; racist view of, 38, 173n83; repatriation of, 105, 151–52; social control over, 67, 152; statistics on numbers of, 2, 73, 74, 108; supervision of, 68–75, 79–80, 81, 89–90, 153, 159; training of, 75, 99–100; wages of, 7, 60, 79, 82–85, 127, 133, 145, 151–52, 167n22; welfare of, 79–80, 108, 128–29; work conditions of, 60, 76, 89, 91, 93, 116, 125, 132, 160; work ethics of, 63, 132. *See also* gang labor system; labor recruitment

Lacton (migrant laborer), 96–97
land grabbing, 28, 29–30
Lawrance, Benjamin N., 9, 135
Leigh, H. W., 152
Lentz, Carola, 67, 79, 153
Liberia: appeal to United States by, 102–3; contract laborers from, 58, 64–70, 161–62, 177–78n53, 178n55; curbing of labor exports from, 103, 105–6, 184n10; French territorial conquest in, 101–2; and labor export, 103, 105–6, 184n8, 184n10; southeastern, 101–2, 184n4
Loring, W. J., 71

Mandingo, 3
Mandorrah, Conah, 94–96
Mariam, Madam, 110–12, 162
Marxism, 11, 81
Master and Servant Ordinance, 65, 66, 70, 92, 114, 161, 163; about, 87–88; amendments to, 137, 143, 181–82n131, 194n135
Maxwell, William Edward, 63
Mende, 56, 107, 127, 185n24
Methodist Missionary Society, 12
Migeod, F. W. H., 5, 110, 119, 136–37; biographical background, 135–36; collaboration with mine management by, 124, 125, 134–35; and Concessions Labor Ordinance, 137, 141; and labor recruitment, 144, 156–57; loyalty to employers by, 134, 191n72; mediating role of, 131, 132–34,

163, 191n65; personal papers of, 12–13, 94; and regulation of labor, 126–27, 128–29, 138–39. *See also* government transport department

Miller, Charles Robert, 90–91

Milner, Alfred, 187n5

Mine Managers' Association, 42, 88, 145, 187n1; formation of, 119–20; and government transport office, 118, 119, 124, 131–32, 134–35; and labor control, 18, 123–24, 137–38, 144; and labor recruitment, 122–23

mining entrepreneurs and managers: Africans as, 23; arrival in Africa of, 29, 37–38, 171n48; and Asian labor, 54, 61–64, 120–24; and colonial government, 2, 18, 118, 119, 124, 131–32, 134–35, 149, 158–59; and control of labor force, 18, 108, 123–24, 137–38, 144; depictions of, 37, 40–43, 46–47, 120–23; hiring of workers by, 23, 57, 65, 66, 69, 91, 120; and indigenous labor agents, 68, 70, 79–80, 108, 115–16; and labor disputes, 113–14, 132–34; labor recruitment requests by, 122–23, 148–49; treatment of workers by, 70–71, 133–34; unscrupulous methods used by, 28–29

missionaries, 12

Moor, G. C., 112

Moshi, 107

Nathan, Matthew, 124, 125, 138, 139, 143

Nigeria, 106–7, 136

Northern Copper Company, 41

Northern Territories: as British protectorate, 146–47; as labor reserve, 107, 145, 146–49; recruitment of labor from, 149–53

Nzemans, 55, 56, 63, 107

Obuase, 1, 33, 125, 189n31; gold mines in, 12, 108, 112

Offin River Company, 132

Osborn, Emily Lynn, 9, 135

Panning, 111. *See also* washing

pass laws, 8, 15, 120, 124, 138–39, 198n59; in South Africa, 137, 192n93

Percy Tarbutt and Co., 42

Phillips, Anne, 5

piecework, 58–60

poaching, 116–18

Poore, Philip, 134

Portuguese, 3, 107

Prestea, 1, 33

Prestea mining company, 37, 131, 132, 134

Quentin, Cecil, 41

racial tensions, 38, 173n83

railway, 35, 154, 185n22, 197–98n46; construction of, 1, 2, 16, 33–34, 37, 131

Rand colony. *See* South Africa

Randlords, 40

Report on the Gold Mines, 28

Report on the Transport Department, 147

Rhodes, Cecil, 40, 41

Rhodesia Copper Company, 41

Roan Antelope, 41

Roberts, Richard L., 9, 135

Rundall, W. H., 48

Russell, J. D., 177–78n53

Salisbury, Lord, 32–33

Sam, Thomas Birch Freeman, 57, 126
Sam, William Edward, 57
Sam, William Edward, Jr., 57, 126
Sekondi, 1–2, 91, 125–26, 148, 189n31
Sierra Leone, 2, 76, 94, 99, 106–7
Simpson, George Gaylord, 136
slavery: abolition of, 5–6, 52, 87, 166n14; and wage labor, 76, 87; as widespread, 59
sleeping sickness, 33
Société des Mines d'Or d'Afrique Occidentale, 20
South Africa, 16, 114–15, 120; labor bureau in, 122, 188n15; pass law in, 137, 192n93; white miners in, 121, 187n5
South African War, 34, 120
speculative investments, 20–21, 28
Stockfeld, Gerhard, 46–47, 90, 126, 148
supervisors, indigenous, 81, 153, 159; abuses by, 75; and cash advances, 89–90; mine managers' relationship to, 79–80; origin of system of, 68–75. *See also* gang leaders
surface mining, 55–56, 58, 83, 107–8, 112, 121
Swanzy, Frank and Andrew, 21, 58
Swanzy Estates, 20, 57, 64
Swindell, Kenneth, 5

Tamsu, 20, 37, 130
Tamsu Gold Mining Company, 20, 92, 127; labor conflicts at, 132–34
Tarbutt, Percy Coventry, 40–43, 46, 120–23
Tarkwa: forest environment of, 72; and gold rush, 12, 38–40; mines in, 47, 48–50, 124, 130, 134–35; and railroad, 1, 33, 35
Tarquah Mining and Exploration Company, 20, 48, 117, 118
Taylor, U., 115
Thomas, Roger, 149, 154
Timmani, 56, 107
To the Gold Coast for Gold (Burton and Cameron), 22–23
Toomer, J. Fletcher, 131, 132, 134
transport office. *See* government transport department
Transvaal Chamber of Mines, 120
tributary labor, 56–58

underground mining, 45, 48, 52, 56, 106, 107 133, 150–51, 159; dangers of, 53, 54, 159; stigma against, 8, 83, 161.
United States, 102–3

Vai people, 56, 107
van Waijenburg, Marlous, 83, 167n22

wage labor, 129–30; and capitalism, 10–11; and slavery, 76, 87; transition to, 7–8, 81
wages: and advance payment system, 151–52; conflicts over, 133; and gang leaders, 129, *130*; rates of, 7, 82–85, 127, 145, 167n22; of women, 60
Wangara, Mana, 117
Wangara, Seiden, 118
Wangara people, 107
War of the Golden Stool (Yaa Asantewaa War), 146
Washing, 58, 59, 60. *See also* panning
Wassa: British protectorate established in, 19, 24; capital investments in, 26–27, 33, 145;

existing scholarship on, 6–7; first and second gold booms in, 16–50; geographic position of, 131; as gold-mining region, 1, 3–4; government assistance to mines in, 119–44; infrastructure in, 8, 33–34; labor recruitment patterns in, 52–100, 101–18; Northern Territories as labor reserve for, 145–57. *See also* Gold Coast Colony

Wassa Gold Coast Mining Company, 20, 22, 23

Watherson, A. E., 147, 150, 198n53

White, John, 91

white miners, 18, 29, *39*, 73, *74*, 171n48; and Chinese laborers, 121; and gold rush, 37–38; as labor recruiters, 112–14; as prone to disease, 29, 71–73, 179–80n88; skilled vs. unskilled, 121, 187–88n9; in South Africa, 187n5

Wodehouse, John, 27–28

women: and family migration, 152–53, 197n40; gold-mining work by, 59–60, 111, 163–64; as labor recruiters, 101, 110–12

work conditions, 60, 76, 91, 125, 160; laborers' discontent with, 89, 93, 116, 132

Yorubas, 56, 82, 107, 185n24

Lightning Source UK Ltd.
Milton Keynes UK
UKHW03n1734110418
320882UK00001B/52/P